CARBIDES

PROPERTIES, PRODUCTION, AND APPLICATIONS

CARBIDES

PROPERTIES, PRODUCTION, AND APPLICATIONS

T. Ya. Kosolapova
Institute of Ferrous Metallurgy
Academy of Sciences of the Ukrainian SSR
Kiev, USSR

With a Foreword by
Herbert S. Kalish
Director of Research and Development
Adamas Carbide Corp.
Kenilworth, N. J.

and

Henry H. Hausner
Adjunct Professor
Polytechnic Institute of Brooklyn
Consulting Engineer

Translated from Russian by
N. B. Vaughan
Formerly Editor of Publications
Institute of Metals, London

PLENUM PRESS · NEW YORK–LONDON · 1971

Tat'yana Yakovlevna Kosolapova was born in 1918 and was graduated as a chemical engineer from the Odessa Industrial Institute in 1940. Subsequently she worked at the Perm sulfuric acid plant as head of the plant laboratory and research engineer in the central laboratory. Since 1946 she has worked in the Institute for Ferrous Metallurgy of the Academy of Sciences of the Ukrainian SSR, one of the departments of which has been expanded into the Institute of Problems in Materials Science of the Academy of Sciences of the Ukrainian SSR. Here she served as head of the analytical chemistry laboratory, chief engineer, and junior scientific associate. In 1961 she was awarded the degree of candidate of chemical sciences and became a senior scientific associate. Since 1965 she has been head of the department for the technology of refractory compounds.

The original Russian text, published by Metallurgiya Press in Moscow in 1968, has been corrected by the author for the present edition. The English translation is published under an agreement with Mezhdunarodnaya Kniga, the Soviet book export agency.

КОСОЛАПОВА Татьяна Яковлевна

КАРБИДЫ

KARBIDY

Library of Congress Catalog Card Number 70-128507

ISBN 978-1-4684-8008-5 ISBN 978-1-4684-8006-1 (eBook)
DOI 10.1007/978-1-4684-8006-1

Softcover reprint of the hardcover 1st edition 1971

A Division of Plenum Publishing Corporation
227 West 17th Street, New York, New York 10011

United Kingdom edition published by Plenum Press, London
A Division of Plenum Publishing Company, Ltd.
Donington House, 30 Norfolk Street, London W.C.2, England

Foreword to the American Edition

In keeping up with the latest international literature, we learned that "Carbides" by T. Ya. Kosolapova had been published in 1968 in the Soviet Union. After having read the book we decided that the information it contained is most valuable and should be made available to the scientists and engineers of the Western World and therefore recommended it, in 1969, for translation into English [1].

The great importance of carbide compounds such as those of tungsten, molybdenum, vanadium, titanium, niobium, silicon, calcium, boron, etc., is well known. It is, however, not always realized how important tungsten carbide and its associated refractory metal carbides are in today's highly technological world. The use of tungsten can best illustrate this point. In 1959 the United States' consumption of tungsten that went into end products was 9,750,000 lb. Of that 3,640,000 lb (37% of the total) was used as carbide for cemented carbide cutting tools, mining tools, and wear parts. Ten years later, in 1969, the total tungsten consumption increased to 13,030,000 lb (34% increase) while the amount that went into carbides in various forms increased to 8,362,000 lb (130% increase) and thus the carbide use accounted for 64% of the total tungsten consumption. The use of tungsten carbide is generally estimated to be increasing at a rate of 10−15% per year so that by 1979 one can expect to see about 20,000,000 lb of tungsten consumed per year for tungsten carbide and it will, then, probably account for at least 70% of the total tungsten utilized.

Commensurate with the increased use of tungsten carbide is that of the common additives tantalum carbide, titanium carbide, and niobium carbide.

The use of cemented titanium carbide, which is still in its infancy (with United States production probably at less than 100,000 lb per year), is expected to increase at a significantly more rapid pace. With it often goes the use of molybdenum carbide. We would not venture a guess as to what the production will amount to by 1979, except to indicate that it will have grown significantly faster than the rapid pace of tungsten carbide.

Other exciting uses for carbide compounds include the high-temperature industrial and space applications. Boron carbide is the major control-rod material for nuclear power reactors. If the trend to nuclear power continues, this carbide too will be produced in large quantities in our advanced technology 10 years hence.

There is no doubt that up to date information on carbide materials is needed.

The first work on refractory metal compounds, a rather thin booklet, was written in German by K. Becker [2] in 1933. Since then many books have been published on this subject, just a few of which will be mentioned.

Schwarzkopf and Kieffer's [3] "Refractory Hard Metals," published in 1953, is usually regarded as the classic work on the subject. The authors, both Austrian, are known as pioneers in the development of carbides and the largest part of the book is, therefore, devoted to carbide materials. Much of it is not out-dated today and contains valuable information even though it was written 20 years ago. It contains basic information on fabrication and properties of carbides and many of the methods of producing carbides described are still widely used. Unfortunately, the book is out of print.

The "Handbook of Hard Metals" [4] by W. Dawihl, another pioneer in the field, was published in 1955. It contains important information on the preparation of carbide powders but more information on the production of hard metal alloys, i.e., sintered materials. This book too is out of print.

Very valuable information on refractory metal compounds and especially on carbides is to be found in "Hartstoffe" [5] by R. Kieffer and F. Benesovsky (Vienna, 1963). This excellent book, however, has not yet been translated from the German, although translation has been recommended.

One of the finest books on intermetallic compounds, "Interstitial Alloys" [6] by H. J. Goldschmidt, was published in 1967 and contains a basic and highly scientific survey on all types of metal compounds and includes extensive data on carbides. It was heretofore the best scientific information available on carbides.

In 1968 "Fundamentals of Refractory Compounds" [7] was published. It is a scientific book edited by H. H. Hausner and M. G. Bowman, and is based on a series of lectures presented at the University of California in Los Angeles.

An important part of development of carbides and other compound metals occurred in the Soviet Union. We call attention to the work of G. A. Meerson, S. I. Bashkirov, V. A. Ivensen, and especially to that of G. V. Samsonov. In 1964 Plenum Press published the "Handbook of High Temperature Materials" with the "Materials Index" [8] by P. T. B. Shaffer and the "Properties Index" [9] by G. V. Samsonov. Today these two books probably represent the most important source of data on properties and applications of refractory compounds. However they do not contain information on the methods of fabrication. In the same year, 1964, S. I. Bashkirov [10] edited a book "Hard Metals Production, Technology, and Research in the USSR." This book deals mainly with the preparation of sintered material but contains limited information on the production of carbides.

Some information on carbides is disclosed in the 1966 book "Coatings of High-Temperature Materials" [11] edited by H. H. Hausner, with chapters by G. V. Samsonov and A. P. Epik, W. A. Gibeaut and E. S. Bartlett, and D. H. Leeds. Coating is a very special and important application of carbides.

Kosolapova's "Carbides" is different from all the abovementioned books. It not only describes in detail the structures and physical and chemical properties of carbides but also discusses extensively the methods for carbide production, and it is up to date. The carbides discussed in the book are organized according to groups in the periodic table and, in the last chapter, extensive data on the use and application of all the carbides are provided. The book covers only the carbides in their uncemented forms, but this is important to great segments of all industry. The final chapter cites six general fields of application:

1. Refractory materials,
2. Corrosion-resistant materials and materials for the chemical industry,
3. Materials for nuclear-energy production,
4. Materials for the manufacture of aircraft and rockets,
5. Materials for electrical and radio applications, and
6. Superhard and wear-resistant materials.

Within these fields also are cited hundreds of specific classes of applications.

While there is much research occurring in new and exciting uses of carbides (such as for catalysts in the petroleum industry), the major usage remains cemented carbides, particularly those based on tungsten carbide and titanium carbide, but this rapidly growing field needs the information imparted in this book. Commercial cemented carbides no longer depend merely on tungsten carbide with a binder and additions of titanium carbide and tantalum carbide. Today commercial use is being made of carbides of molybdenum, niobium, chromium, and vanadium. Research is taking place which will surely see the early incorporation of additions of carbides of hafnium and zirconium, and, later, the use of carbides of rare earths and alkali metals. The information imparted in this book will be of great assistance to those researchers working in the field of cemented carbides for machining tools and wear applications as well as those involved in the other numerous application areas cited.

Miss Kosolapova has closely cooperated with Prof. G. V. Samsonov and other Russian scientists in writing her manuscript. In the well-illustrated text 921 references are reviewed, approximately 50% of which cover the literature in English and the other 50% those in Russian.

This important reference work is a credit to an outstanding Russian scientist. The translator has performed a creditable service to the author and to those fortunate enough to have this fine work available in the English language.

H. S. Kalish,
Director of Research and Development,
Adamas Carbide Corp.

H. H. Hausner,
Adj. Professor,
Polytechnic Institute of Brooklyn,
Consulting Engineer

References

1. Hoeganaes P/M Newsletter (H. H. Hausner, ed.), Vol. 3, No. 1 (January 1969).
2. K. Becker, Hochschmelzende Hartstoffe und Ihre technische Anwendung, Verlag Chemie, Berlin (1933).
3. P. Schwarzkopf and R. Kieffer, Refractory Hard Metals, MacMillan Company, New York (1953).
4. W. Dawihl, A Handbook of Hard Metals, Her Majesty's Stationary Office, London (1955).
5. R. Kieffer and F. Benesovsky, Hartstoffe, Springer Verlag, Vienna (1963).
6. H. J. Goldschmidt, Interstitial Alloys, Plenum Press, New York (1967).
7. H. H. Hausner and M. G. Bowman (ed.), Fundamentals of Refractory Compounds, Plenum Press, New York (1968).
8. P. T. B. Shaffer, High Temperature Materials, Vol. 1, Materials Index, Plenum Press, New York (1964).
9. G. V. Samsonov, High Temperature Materials, Vol. 2, Properties Index, Plenum Press, New York (1964).
10. S. I. Bashkirov (ed.), Hard Metals Production, Technology and Research in the USSR, MacMillan Company, New York (1964).
11. H. H. Hausner (ed.), Coatings of High-Temperature Materials, Plenum Press, New York (1966).

Foreword to the Russian Edition

Carbides belong to a broad class of compounds having a number of valuable properties which make them promising materials for use in various new fields of technology. Metal carbides, particularly carbides of the transition metals, possess high melting points and great hardness, high chemical resistance, electrical and thermal conductivities of a metallic character, and also a number of special properties such as a capacity to be transformed to the superconducting state at relatively high temperatures, high emission properties, etc. In spite of the large numbers of researches devoted to the study of the properties of carbides and methods of producing them, carbides have not yet been adequately investigated, and considerable discrepancies are often found among the data reported in different papers. Moreover, these data are scattered among numerous journals, and this makes it difficult to obtain a thorough acquaintance with this class of compound.

A detailed investigation of the properties of carbides is of both practical importance and scientific interest from the point of view of elucidating the nature of the chemical bond in carbides and the dependence of the properties on the nature of this bond, and of discovering ways in which materials with the above properties may be produced.

The present book contains information on all the known Me — C systems, on the methods of obtaining the individual phases, and on their physical and chemical properties.

The author is deeply grateful to Corresponding Member of the Academy of Sciences of the Ukrainian SSR G. V. Samsonov for valuable suggestions and constant help during the writing of the book, to Professor P. V. Gel'd for reading the manuscript and making valuable comments, and also to her colleagues in the Laboratory for the Technology of Inorganic Compounds in the Institute for Problems in Materials Science of the Academy of Sciences of the Ukrainian SSR, in particular, to G. N. Makarenko, V. B. Fedorus, O. F. Kvas, and A. V. Tkachenko for assistance in planning the book and reading the manuscript.

Contents

Chapter I

The Structure and Physicochemical Properties of Carbides

Structures

The structural types of carbides can be divided into those with isolated carbon atoms, those with isolated pairs of carbon atoms, those with chains of carbon atoms, and those with networks of carbon atoms [11].

The alkali metals form carbides in which the structural elements consisting of carbon atoms become more complex on passing from lithium carbide to the carbides of potassium, sodium, rubidium, and cesium. Thus lithium forms a single carbide Li_2C_2 [4], while sodium forms NaC_8, NaC_{16}, and NaC_{64}, in addition to Na_2C_2 [5]; for potassium the most typical carbides are KC_8, KC_{16}, KC_{24}, KC_{36}, KC_{48}, KC_{60}, and similarly for rubidium and cesium, except that the carbide Cs_2C_2 is not readily formed. These carbides have graphite-like lattices in which the metal atoms are situated between the layers of carbon atoms (Fig. 1). The most likely arrangement of the metal atoms is that in which their projection on the carbon networks gives a distribution of them at the centers of the carbon hexagons [895] (Fig. 2). The distances between the carbon networks are 0.540 nm for KC_8, KC_{24}, and KC_{36}; 0.565 nm for RbC_8, RbC_{24}, and RbC_{36}; 0.460 nm for NaC_{64}; and 0.594 nm for CsC_8.*

* Throughout the text the lattice parameters are given in nanometers (1 nm = 10 Å).

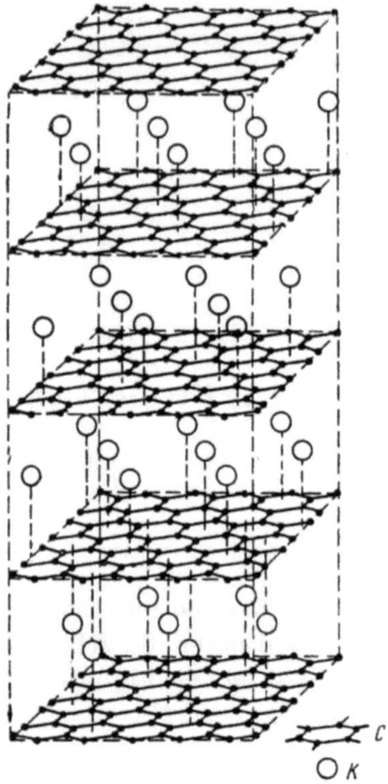

Fig. 1. The structure of potassium c bide KC_8.

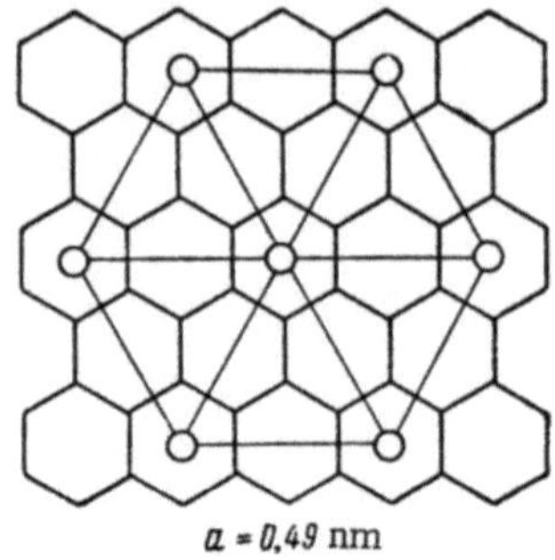

Fig. 2. Arrangement of alkali-metal atoms in a compound of graphite with potassium in relation to the hexagonal carbon networks.

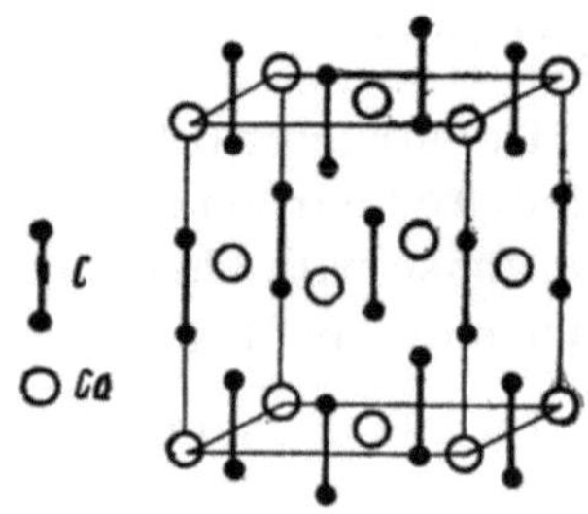

Fig. 3. The structural type of CaC_2.

On passing to carbides of the alkaline-earth metals, the tendency to the formation of complex anions from carbon atoms diminishes, and they are characterized by MeC_2-type carbide phases having structures consisting of isolated pairs of carbon atoms. The carbides CaC_2, SrC_2, and BaC_2 crystallize in the CaC_2-type tetragonal cell [896] (Fig. 3). The carbide MgC_2 crystallizes in the ThC_2-type tetragonal cell (Fig. 4).

Lanthanides and also scandium and yttrium, which lie close to them in the periodic system, form carbide phases corresponding in composition to Me_3C, MeC, Me_2C_3, and MeC_2.

Yttrium and all the rare-earth metals except lanthanum, cerium, praseodymium, and neodymium form carbides of composition Me_3C. These phases crystallize in the Fe_4N-type cubic structure,

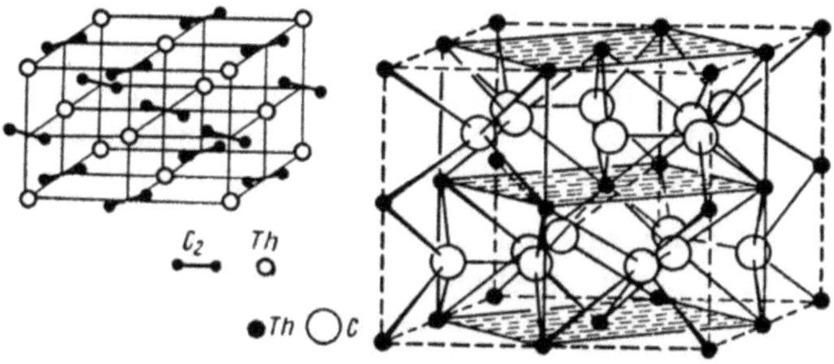

Fig. 4. The structural type of ThC_2.

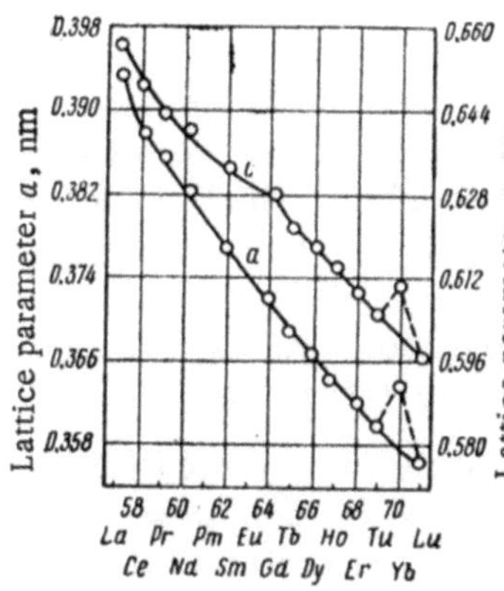

Fig. 5. Dependence of the lattice parameters of dicarbides on the atomic number of the metal in them.

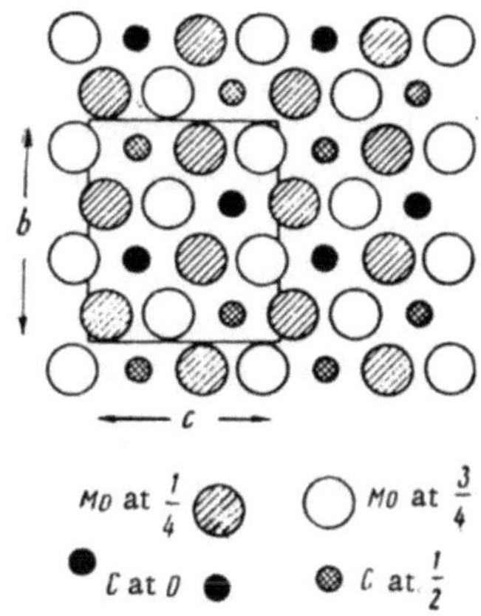

Fig. 6. Arrangement of carbon and molybdenum atoms in the Mo_2C lattice.

which is similar to the NaCl-type fcc lattice, but with a deficiency of carbon atoms.

MeC phases have been found for yttrium, scandium, and cerium; they crystallize in the NaCl-type fcc lattice [85, 119, 185, 186, 820].

The sesquicarbides Me_2C_3 crystallize mainly in the Pu_2C_3-type bcc structure (from La_2C_3 to Ho_2C_3). The structures of the sesquicarbides of yttrium, erbium, thulium, and lutecium have not yet been finally determined. Yttrium and all the lanthanides form dicarbides which crystallize in the CaC_2-type bc-tetragonal structure (Fig. 3) with two atoms per unit cell. The dependence of the lattice parameters of the dicarbides on the atomic number of the metal (Fig. 5) is evidence of the lanthanide contraction effect which has been observed in other carbide phases of rare-earth elements.

A study of the phase diagrams of the rare-earth metal (Gd, Dy, Ho, and Er) — carbon systems in the composition range $MeC_{0.3}-MeC_{1.0}$ has led to the discovery of a new phase near to $MeC_{0.50}$ [930], based on a fcc cell. Thus the $MeC_{0.50}$ phases crystallize in a fcc lattice. With increase in carbon content, a rhombohedral distortion appears, which reaches a maximum at $Me_2C_{0.5}$ and begins to disappear in compounds close to $MeC_{0.6}$.

Carbides of the actinides are similar in structure to those of the rare-earth elements. They form the phases MeC, Me_2C_3, and MeC_2. The monocarbides crystallize in the NaCl-type fcc lattice. The sesquicarbides (except in the case of thorium, which does not form such a carbide) crystallize in the Pu_2C_3-type bcc lattice, in which the carbon atoms form pairs. The C — C distance is 0.154 nm. Dicarbides are known for uranium and thorium; they crystallize in the ThC_2 (see Fig. 4) and CaC_2 (UC_2) fc-tetragonal lattices.

Carbides of the transition metals mainly form interstitial phases or phases similar to them with isolated carbon atoms and with structural chains of carbon atoms [11]. According to Hagg's rule, interstitial-phase structures are formed when the ratio of the radius of the carbon atom to that of the metal atom does not exceed 0.59.

The carbon atoms are inserted into the octahedral or tetrahedral spaces between the metal atoms, the initial metal lattice being usually transformed into a lattice of high symmetry [12].

Metals of subgroup IVa form monocarbides which crystallize in the NaCl-type fcc lattice. These carbides are characterized by a wide homogeneity range. The lattice parameters of these carbides increase linearly with rise in carbon content.

Metals of subgroup Va form carbide phases Me_2C_{1-x} and MeC_{1-x}. The former crystallize in the Mo_2C-type hexagonal lattice consisting of close-packed hexagonal molybdenum layers with carbon atoms situated in the octahedral spaces [550] (Fig. 6). The monocarbides crystallize in the NaCl-type fcc lattice. The homo-

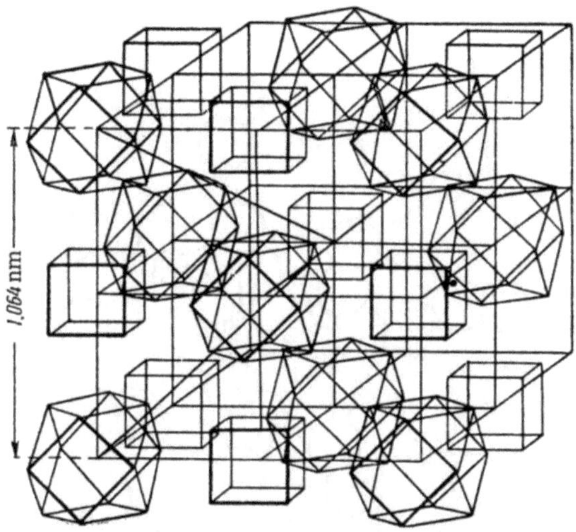

Fig. 7. The structural type of $Cr_{23}C_6$.

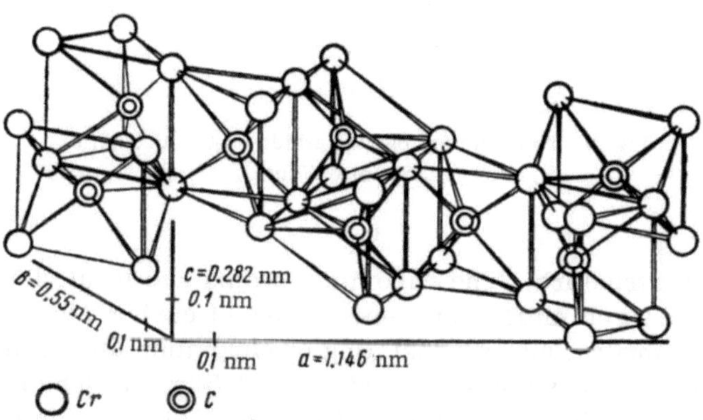

Fig. 8. The structure of chromium carbide Cr_3C_2.

geneity ranges of the monocarbides of metals of subgroup Va are somewhat narrower than those of the monocarbides of metals of subgroup IVa and wider than those of the sesquicarbides of metals of subgroup Va.

Of the chromium carbide phases ($Cr_{23}C_6$, Cr_7C_3, and Cr_3C_2) the lowest in carbon content ($Cr_{23}C_6$) has a structure with isolated carbon atoms; it crystallizes in the fcc lattice and has an individu-

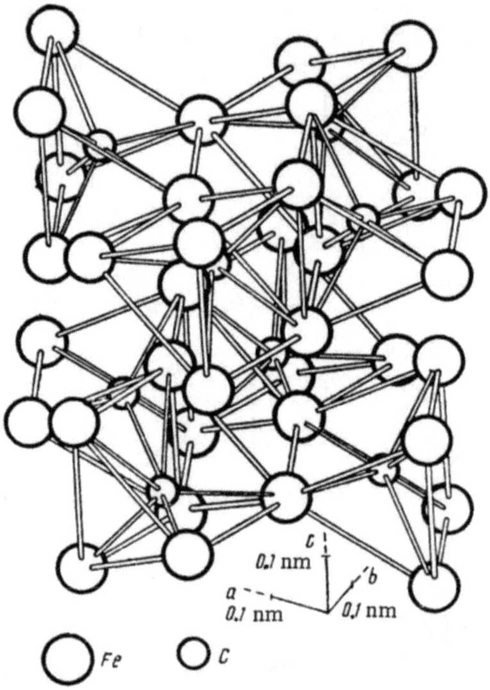

Fig. 9. The structure of the carbide Fe_3C. a) 0.45; b) 0.50; c) 0.67 nm.

al type of structure. The chromium atoms are located at the centers of the Archimidean cubo-octahedra of carbon atoms and truncated tetrahedra; cubes consisting of chromium atoms are located within the Fedorov cubo-octahedra [896] (Fig. 7). The higher chromium carbides form types of structure having chains of carbon atoms. The carbide Cr_7C_3 crystallizes in a hexagonal lattice similar in type to the structure of Ru_7B_3 [897].

The carbide Cr_3C_2 has a rhombic structure of an individual type, in which each of the eight carbon atoms is situated at the center of a triangular prism. The corners of the prism are occupied by chromium atoms (Fig. 8). The shortest distance Cr — Cr is 0.2487 nm and that for Cr — C, 0.201 nm [799].

Molybdenum and tungsten form Me_2C_{1-x} and MeC phases. The former phases crystallize in the Mo_2C-type hexagonal lattice and the latter in the MoC-type hexagonal lattice.

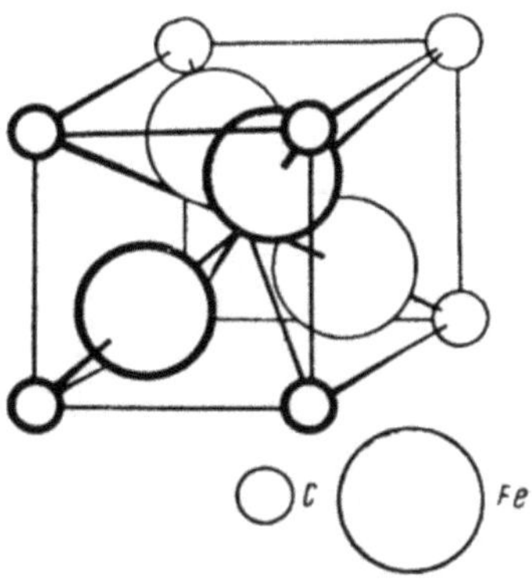

Fig. 10. The structure of the carbide Fe_4C.

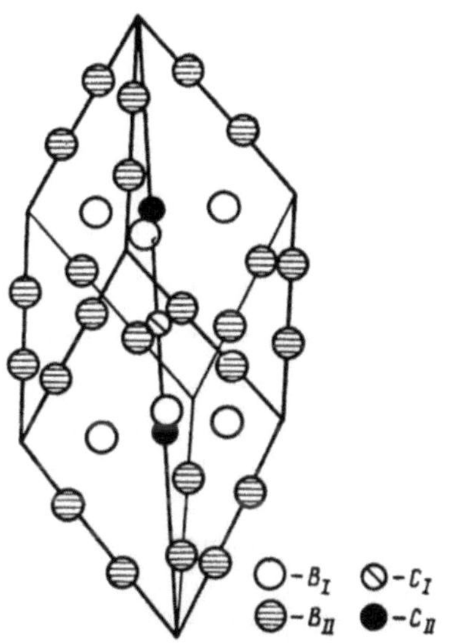

Fig. 11. The structure of the carbide B_4C.

The homogeneity range of carbides of subgroup VIa is considerably narrower than that of carbides of metals of subgroups IVa and Va. In carbides of higher carbon content the range practically disappears.

Among the metals of subgroup VIIa, manganese is known to have the following carbide phases: $Mn_{23}C_6$, $Mn_{15}C_4$, Mn_5C_2, Mn_7C_3, and Mn_3C. The last crystallizes in an Fe_3C-type rhombic cell, in which the metal atoms form triangular prisms with carbon atoms located at the center [663] (Fig. 9). The carbides $Mn_{23}C_6$ and Mn_7C_3 crystallize in cubic and hexagonal lattices of the $Cr_{23}C_6$ and Cr_7C_3 types, respectively.

Metals of group VIII form the carbide phases Me_4C (Fe_4C), Me_3C (Fe_3C, Co_3C, and Ni_3C), and Me_2C (Fe_2C and Co_2C). The structure of Fe_4C is shown in Fig. 10. The carbides Me_3C crystallize in the Fe_3C-type rhombic lattice (Fig. 9) and Me_2C in the hexagonal lattice. The homogeneity ranges of carbides of metals of subgroup VIIa and group VIII are very narrow, extending to not more than 1-2 at.% C. Thus, the width of the homogeneity range diminishes on passing from the carbides of transition metals of group IV to those of transition metals of groups V, VI, VII, and VIII. According to data in [13, 14, 172], the width of the homogeneity range of carbides of the transition metals of groups IV-VI of the periodic system depends on the statistical weight of the stable electron

TABLE 1. The Structures and Crystallo-

Phase*	r_C/r_{Me}	Homogeneity range at. %	Homogeneity range wt. %	Structure	Structural elements from C atoms	Lattice type
1	2	3	4	5	6	7
Li_2C_2	0.50	—	—	Rhombic	—	—
Na_2C_2	0.48	—	—	Hexagonal	—	—
NaC_8	0.43	—	—	"	Structural networks	KC_8
NaC_{16}	0.43	—	—	"	The same	KC_{16}
NaC_{64}	0.43	—	—	"	" "	—
K_2C_2	0.33	—	—	"	" "	—
KC_8	0.33	—	—	"	" "	KC_8
KC_{16}	0.33	—	—	"	" "	KC_{16}
KC_{24}	0.33	—	—	"	" "	—
KC_{36}	0.33	—	—	"	" "	—
KC_{48}	0.33	—	—	"	" "	—
KC_{60}	0.33	—	—	"	" "	—
RbC_8	0.31	—	—	"	" "	KC_8
RbC_{16}	0.31	—	—	"	" "	KC_{16}
RbC_{24}	0.31	—	—	"	" "	—
RbC_{36}	0.31	—	—	"	" "	—
RbC_{48}	0.31	—	—	"	" "	—
RbC_{60}	0.31	—	—	"	" "	—
CsC_8	0.29	—	—	"	" "	KC_8
CsC_{16}	0.29	—	—	"	" "	KC_{16}
CsC_{24}	0.29	—	—	"	" "	—
CsC_{36}	0.29	—	—	"	" "	—
CsC_{48}	0.29	—	—	"	" "	—
CsC_{60}	0.29	—	—	"	" "	—
Be_2C	0.68	—	—	Cubic	—	CaF_2
BeC_2	0.68	—	—	—	—	—
Mg_2C_3	0.48	—	—	Hexagonal	—	—
MgC_2	0.48	—	—	Tetragonal	Isolated pairs	ThC_2
CaC_2	0.39	—	—	"	The same	CaC_2
SrC_2	0.36	—	—	Tetragonal	Isolated pairs	CaC_2

* Phases having a homogeneity range are designated MeC_{1-x} In those given. In phases whose composition is close to the stoichiometric, x is

chemical Properties of Carbides

Space group	Lattice parameters, nm			Density, g/cm³		References
	a	b	c	x-ray	pycno-metric	
8	9	10	11	12	13	14
—	0.3655	0.5440	0.4830	1.31	1.30	[241]
—	0.6756	—	1.2688	1.60	1.58	[72]
—	—	—	—	—	—	[5]
—	—	—	—	—	—	[5]
—	—	—	—	—	—	
D_{4h}^{20}	0,758	—	1.469	1.62	—	[72]
—	0.494	—	2.1345	—	—	[895, 896]
—	0.494	—	1.745	—	—	
—	—	—	—	—	—	
—	—	—	—	—	—	
—	—	—	—	—	—	
—	—	—	—	—	—	
—	0.494	—	2.273	—	—	[896]
—	0.494	—	1.795	—	—	
—	—	—	—	—	—	[10,895]
—	—	—	—	—	—	
—	—	—	—	—	—	[10,895]
—	—	—	—	—	—	
—	0.494	—	2.376	—	—	
—	0.494	—	1.851	—	—	
—	—	—	—	—	—	
—	—	—	—	—	—	
—	—	—	—	—	—	
—	—	—	—	—	—	
$Fm3m(O_h^5)$	0.434	—	—	2.437	2.26	[89, 93, 97]
—	—	—	—	—	—	—
—	0.745	—	1.061	—	—	[85]
14 *mmm* (D_{4h}^{14})	0.555	—	0.508	2.07	—	[896]
14/*mmm* (D_{4h}^{17})	0.548	—	0.637	2.21	2.10	[119, 896]
14/*mmm* (D_{4h}^{17})	0.4121	—	0.6755	3.26	3.19	[93, 933]

cases where the composition of the phase has been determined, the values of x are near 0. This applies to data given throughout the following tables and text.

TABLE 1

Phase	r_C/r_{Me}	Homogeneity range at.%	Homogeneity range wt.%	Structure	Structural elements from C atoms	Lattice type
1	2	3	4	5	6	7
BaC_2	0.35	—	—	Tetrag.	Isolated pairs	CaC_2
ScC	0.47	—	—	Cubic	Isolated atoms	NaCl
$Y_3C_{0.9}$	0.43	—	—	"	The same	Fe_4N
YC	0.43	—	—	"	" "	NaCl
Y_2C_3	0.43	—	—	—	—	—
YC_2	0.43	—	—	Tetrag.	Isolated pairs	CaC_2
La_2C_{3-x}	0.41	56.2—60.0	9.98—11.58	Cubic	The same	Pu_2C_3
α-LaC_2	0.41	—	—	Tetrag.	" "	CaC_2
β-LaC_2	0.41	—	—	Cubic	" "	FeS_2
Ce_2C_3	0.42	—	—	"	Isolated pairs	Pu_2C_3
CeC_2	0.42	—	—	Tetrag.	The same	CaC_2
Pr_2C_3	0.42	—	—	Cubic	" "	Pu_2C_3

(continued)

Space group	Lattice parameters, nm			Density, g/cm^3		References
	a	*b*	*c*	x-ray	pycnomet.	
8	9	10	11	12	13	14
$I4/mmm$ (D_{4h}^{17})	0.4396	—	0.7109	3.72	3.90	[93, 119, 933]
$Fm3m$ (O_h^5)	0.499	—	—	4.05	3.60	[179—181]
$Pm3m$ (O_h^1)	0.5068	—	—	5.41	4.73	[7, 864]
$Fm3m$ (O_h^5)	0.499	—	—	5.38	—	[6]
—	—	—	—	—	3.66	
$I4/mmm$ (D_{4h}^{17})	0.366	—	0.6169	4.58	3.95	[7,93]
$I\bar{4}3d$ (T_d^6)	0.88185 $x\approx0$	—	—	6.07	—	
$I4/mmm$ (D_{4h}^{17})	0.3934	—	0.6572	5.35	5.29	[6, 7]
$Fm3m$ (O_h^5)	0.60	—	—	5.0	—	
$I\bar{4}3d$ (T_d^6)	0.84476	—	—	6.969	—	[7]
$I4/mmm$ (D_{4h}^{17})	0.3878	—	0.6488	5.56	5.56	[7, 896]
$I\bar{4}3d$ (T_d^6)	0.86072	—	—	6.621	—	[7]

TABLE 1

Phase	r_C/r_{Me}	Homogeneity range		Structure	Structural elements from C atoms	Lattice type
		at. %	wt. %			
1	2	3	4	5	6	7
PrC_2	0.42	—	—	Tetrag.	Isolated pairs	CaC_2
Nd_2C_3	0.42	—	—	Cubic	The same	Pu_2C_3
NdC_2	0.42	—	—	Tetrag.	" "	CaC_2
Sm_3C	0.42	—	—	Cubic	Isolated atoms	Fe_4N
Sm_2C_3	0.42	—	—	"	Isolated pairs	Pu_2C_3
SmC_2	0.42	—	—	Tetrag.	The same	CaC_2
Gd_3C	0.38	—	—	Cubic	Isolated atoms	Fe_4N
Gd_2C_3	0.38	—	—	"	Isolated pairs	Pu_2C_3
GdC_2	0.38	—	—	Tetrag.	The same	CaC_2
Tb_3C	0.44	—	—	Cubic	Isolated atoms	Fe_4N
Tb_2C_3	0.44	—	—	"	Isolated pairs	Pu_2C_3
TbC_2	0.44	—	—	Tetrag.	The same	CaC_2
Dy_3C	0.44	—	—	Cubic	Isolated atoms	Fe_4N

(continued)

Space group	Lattice parameters, nm			Density, g/cm³		References
	a	*b*	*c*	x-ray	pycnomet.	
8	9	10	11	12	13	14
$I4/mmm$ (D_{4h}^{17})	0.3855	—	0.6434	5.728	5.58	[7]
$I\bar{4}3d$ (T_d^6)	0.85478	—	—	6.902	—	
$I4/mmm$ (D_{4h}^{17})	0.3823	—	0.6405	5,970	5.93	
$Pm3m$ (O_h^1)	0.5172	—	—	8.139	—	[6,7]
$I\bar{4}3d$ (T_d^6)	0.84257	—	—	7.477	—	
$I4/mmm$ (D_{4h}^{17})	0.3770	—	0.6331	6.50	6.421	
$Pm3m$ (O_h^1)	0.5126	—	—	8.701	—	
$I\bar{4}3d$ (T_h^6)	0.83407	—	—	8.024	—	
$I4/mmm$ (D_{4h}^{17})	0.3718	—	0.6275	6.939	6.93	
$Pm3m$ (O_h^1)	0.5107	—	—	8.882	—	
$I\bar{4}3d$ (T_d^6)	0.82617	—	—	8.335	—	
$I4/mmm$ (D_{4h}^{17})	0.3690	—	0.6217	7.176	7.09	
$Pm3m$ (O_h^1)	0.5079	—	—	9.211	—	

TABLE 1

Phase	r_C/r_{Me}	Homogeneity range		Structure	Structural elements from C atoms	Lattice type
		at. %	wt. %			
1	2	3	4	5	6	7
Dy_2C_3	0.44	—	—	Cubic	Isolated pairs	Pu_2C_3
DyC_2	0.44	—	—	Tetrag.	The same	CaC_2
Ho_3C	0.44	—	—	Cubic	Isolated atoms	Fe_4N
Ho_2C_3	0.44	—	—	"	Isolated pairs	Pu_2C_3
HoC_2	0.44	—	—	Tetrag.	The same	CaC_2
Er_3C	0.44	—	—	Cubic	Isolated atoms	Fe_4N
ErC_2	0.44	—	—	Tetrag.	Isolated pairs	CaC_2
Tm_3C	0.44	—	—	Cubic	Isolated atoms	Fe_4N
TmC_2	0.44	—	—	Tetrag.	Isolated pairs	CaC_2
Yb_3C	0.40	—	—	Cubic	Isolated atoms	Fe_4N
YbC_2	0.40	—	—	Tetrag.	Isolated pairs	CaC_2
Lu_3C	0.44	—	—	Cubic	Isolated atoms	Fe_4N

(continued)

Space group	Lattice parameters, nm			Density, g/cm^3		References
	a	*b*	*c*	x-ray	pycnomet.	
8	9	10	11	12	13	14
$I\bar{4}3d$ (T_d^6)	0.8198	—	—	—	—	
$I4/mmm$ (D_{4h}^{17})	0.3669	—	0.6176	7.450	7.38	
$Pm3m$ (O_h^1)	0.5061	—	—	9.434	—	
$I\bar{4}3d$ (T_d^6)	0.8176	—	—	8.892	—	
$I4/mmm$ (D_{4h}^{17})	0.3643	—	0.6139	7.701	7.76	
$Pm3m$ (O_h^1)	0.5034	—	—	4.708	—	
$I4/mmm$ (D_{4h}^{17})	0.3620	—	0.6094	7.954	7.70	[6,7]
$Pm3m$ (O_h^1)	0.5016	—	—	9.901	—	
$I4/mmm$ (D_{4h}^{17})	0.360	—	0.6047	8.175	—	
$Pm3m$ (O_h^1)	0.4993	—	—	10.26	—	
$I4/mmm$ (D_{4h}^{17})	0.3637	—	0.6109	8.097	7.97	
$Pm3m$ (O_h^1)	0.4965	—	—	10.54	—	

TABLE 1

Phase	r_C/r_{Me}	Homogeneity range		Structure	Structural elements from C atoms	Lattice type
		at. %	wt. %			
1	2	3	4	5	6	7
LuC_2	0.44	—	—	Tetrag	Isolated pairs	CaC_2
ThC_{1-x} $0.05<x<0.1$	0.43	—	—	Cubic	Isolated atoms	NaCl
ThC_{2-x} $x=0.01$	0.43	—	8—9.56	Monoclinic	Isolated pairs	—
UC_{1-x}	0.50	—	—	Cubic	Isolated atoms	NaCl
U_2C_3	0.50	—	—	"	Isolated pairs	Pu_2C_3
α-UC_{2-x}	0.50	—	—	Tetrag.	The same	CaC_2
β-UC_{2-x}	0.50	—	—	Cubic		CaF_2
NpC	—	—	—	"	Isolated atoms	NaCl
Np_2C_3	—	—	—	"	Isolated pairs	Pu_2C_3
PuC_{1-x} $x=0.15$	0.48	46—48	4.04—4.34	"	Isolated atoms	NaCl
Pu_2C_3	0.48	—	—	"	Isolated pairs	Pu_2C_3
TiC_{1-x} $x=0.079$	0.53	37—50	13—20	"	Isolated atoms	NaCl
ZrC_{1-x} $x=0.043$	0.48	38.5—50	7.6—11.64	"	The same	NaCl
HfC_{1-x} $x=0.03$	0.48	36—50	3.7—6.3	"	" "	NaCl

(continued)

Space group	Lattice parameters, nm			Density, g/cm^3		References
	a	b	c	x-ray	pycnomet.	
8	9	10	11	12	13	14
I4/mmm (D_{4h}^{17})	0.3563	—	0.5964	8.728	—	[6,7]
Fm3m (O_h^5)	0.5335	—	—	10.61	—	[197, 208]
—	0.6691	0.4231	0.6744	9.6	—	[134, 200, 215]
Fm3m (O_h^5)	0.4959	—	—	13.63	12.97	[199, 264]
$I\bar{4}3d$ (T_d^6)	0.8088	—	—	12.88	12.7	[227]
I4/mmm (D_{4h}^{17})	0.3517	—	0.5987	11.79	11.28	[259]
Fm3m (O_h^5)	0.546	—	—	—	—	[221, 250]
Fm3m (O_h^5)	0.50026	—	—	—	—	[936]
$I\bar{4}3d$ (T_d^6)	0.81023	—	—	12.785	—	[936]
Fm3m (O_h^5)	0.49582	—	—	13.42	—	[265]
$I\bar{4}3d$ (T_d^6)	0.81258	—	—	12.70	—	[269]
Fm3m (O_h^5)	0.43178	—	—	4.92	4.93	[286, 335]
Fm3m (O_h^5)	0.46828	—	—	6.66	6.73	[335, 374]
Fm3m (O_h^5)	0.46395	—	—	12.67	12.60	[407]

TABLE 1

Phase	r_C/r_{Me}	Homogeneity range		Structure	Structural elements from C atoms	Lattice type
		at.%	wt.%			
1	2	3	4	5	6	7
V_2C_{1-x} $x \approx 0$	0.57	31—33.3	9.9—10.5	Hexag.	Isolated atoms	Mo_2C
VC_{1-x} $x=0.12$	0.57	40—47	13.4—17.2	Cubic	The same	NaCl
Nb_2C_{1-x} $x \approx 0$	0.53	31.0—33.3	5.8—6.1	Hexag.	" "	Mo_2C
NbC_{1-x} $x=0.116$	0.53	41.2—50.0	8.3—11.4	Cubic	" "	NaCl
Ta_2C_{1-x} $x \approx 0$	0.53	30.3—33.3	2.8—3.2	Hexag.	" "	Mo_2C
TaC_{1-x} $x=0.09$	0.53	42.2—49	4.6—6.0	Cubic	" "	NaCl
$Cr_{23}C_6$	0.61	—	—	"	" "	$Cr_{23}C_6$
Cr_7C_3	0.61	—	—	Hexag.	Chains of atoms	Cr_7C_3
Cr_3C_2	0.61	—	—	Rhombic	The same	Cr_3C_2
Mo_2C_{1-x} $x \approx 0$	0.55	31.2—33.3	5.4—5.9	Hexag.	Isolated atoms	Mo_2C
W_2C_{1-x} $x \approx 0$	0.55	29.5—33.3	2.7—3.2	"	The same	Mo_2C
WC	0.55	—	—	"	" "	MoC
$Mn_{23}C_6$	0.59	—	—	Cubic	" "	$Cr_{23}C_6$
$Mn_{15}C_4$	0.59	—	—	Hexag.	—	—

(continued)

Space group	Lattice parameters, nm			Density, g/cm³		References
	a	*b*	*c*	x-ray	pycnomet.	
8	9	10	11	12	13	14
$C\bar{3}m$ (D_{3d}^{3})	0.290	—	0.4587	5.75	—	[419]
$Fm3m$ (O_h^5)	0.4159	—	—	5.48	5.36	[419, 864]
$C\bar{3}m$ (D_{3d}^{3})	0.3122	—	0.4964	7.85	7.86	[442, 444]
$Fm3m$ (O_h^5)	0.44691	—	—	7.82	7.56	[335, 442, 444]
$C\bar{3}m$ (D_{3d}^{3})	0.3104	—	0.4941	15.05	14.8	[466, 468]
$Fm3m$ (O_h^5)	0.4456	—	—	14.4	14.3	[468, 469]
$Fm3m$ (O_h^5)	1.0638	—	—	6.98	6.97	
$P6mc$ (C_{6v}^{4})	1.398	—	0.4523	6.97	6.92	[371, 490]
$Pbnm$ (D_{2h}^{16})	0.282	0.552	1.146	6.74	6.68	
$C\bar{3}m$ (D_{3d}^{3})	0.3002	—	0.4729	9.06	—	[534]
$C\bar{3}m$ (D_{3d}^{3})	0.30008	—	0.47357	17.34	17.2	[65, 152]
$C\bar{6}m2$ (D_{3h}^{1})	0.2903	—	0.2833	15.77	15.5—15.7	[259, 571]
$Fm3m$ (O_h^5)	1.061	—	—	7.53	—	[856]
—	0.7492	—	1.2071	7.30	—	[782, 856]

TABLE 1

Phase	r_C/r_{Me}	Homogeneity range		Structure	Structural elements from C atoms	Lattice type
		at. %	wt. %			
1	2	3	4	5	6	7
Mn_3C	0.59	—	—	Rhombic	Isolated atoms	Fe_3C
Mn_5C_2	0.59	—	—	Monoclinic	—	—
Mn_7C_3	0.59	—	—	Hexag.	Chain of atoms	Cr_7C_3
Fe_3C	0.61	—	—	Rhombic	Isolated atoms	Fe_3C
Fe_2C	0.61	—	—	Hexag	The same	—
Co_3C	0.62	—	—	Rhombic	" "	Fe_3C
Co_2C	0.62	—	—	"	" "	—
Ni_3C	0.62	—	—	"	" "	Fe_3C
Al_4C_3	0.53	—	—	Rhombo-hedral	—	—
$B_{6.5}C$	0.85	—	—	"	—	B_4C
B_4C_{1-x}	0.85	17.6—29.5	19.2—31.7	"	—	B_4C
β-SiC	0.57	—	—	Cubic	—	ZnS

(continued)

Space group	Lattice parameters, nm			Density, g/cm³		References
	a	b	c	x-ray	pycnomet.	
8	9	10	11	12	13	14
Pbmn (D_{2h}^{16})	0.4530	0.5080	0,6772	—	6.89	[11, 856]
—	0.5086	0.4578	1.166	—	—	[580, 581]
Pbmc (C_{6v}^{4})	1.330	—	0.454	3.69	—	[580]
Pbnm (D_{2h}^{16})	0.45235	0,50890	0.67353	7.69	7.07	[856]
—	0.2757	—	0.4346	7.16	—	[152]
Pbnm (D_{2h}^{16})	0.453	0,509	0,674	8.07	—	[152, 856]
—	0.2885	0.4454	0.4364	7.67	—	[660, 670]
Pbnm (D_{2h}^{16})	0.2646	—	0.4329	7.88	7.96	[681]
—	0.853	0.4203	2.584	2.95	2.99	[65, 161]
$R\overline{3}m$ (D_{3d}^{5})	0.563	—	1.219	2.461	—	[685]
$R\overline{3}m$ (D_{3d}^{5})	0.5598.	—	1.212	3.52	3.50	[685, 687]
$F\overline{4}3$ (T_d^2)	0.4358	—	—	3.217	3.211	[856]

configurations of the metal and carbon atoms. The metals under consideration are characterized by the formation of stable d^0, d^5, and d^{10} electron configurations; for carbon, the most stable one is the configuration sp^3, the formation of which is made possible by the transfer of s-electrons to the p-state.

In considering compounds of carbon with nonmetals, the greatest interest attaches to those having such properties as a high melting point, great hardness, abrasive and chemical resistance, and semiconducting properties. Such compounds are the carbides of boron and silicon, which are also discussed in the present book.

The boron compounds B_4C and $B_{6.5}C$ crystallize in a rhombohedral lattice, and, according to [898], B_4C has the structure shown in Fig. 11. Silicon carbide SiC exists in two main modifications: cubic (β-SiC), having the sphalerite type of structure, and α-SiC, which forms a large number of types of structure having hexagonal and rhombohedral lattices [816-818].

The structures and crystallochemical properties of the carbides are given in Table 1.

Thermodynamic and Thermophysical Properties

Whereas carbides of the alkali metals are relatively unstable and readily decompose even at temperatures of the order of 800°C, on passing to carbides of metals of subgroups IVa, Va, and VIa, we observe a large rise in the melting points, which reach a maximum in carbides of metals of subgroups IVa and Va (HfC, NbC, TaC). The melting points of the carbides are usually higher than those of the corresponding metals. Thus, $T_{mp\ Me}/T_{mp\ MeC}$ lies in the range 0.55-0.98.

Mass-spectrometric investigations of the evaporation of the dicarbides of the rare-earth metals have shown that in the vapors over these carbides there are molecules of the dicarbide as well as metal atoms, while in the case of the dicarbides of cerium, praseodymium, and holmium molecules of the tetracarbides have also been found [178, 226].

It has been shown in [337] that the carbides of titanium, zirconium, and hafnium evaporate without any change in composition in the vapors. On evaporation in the temperature range 2770-3300°K, the carbides of niobium and tantalum change in composition, carbon evaporating preferentially to give the composition $MeC_{0.7}$ — $MeC_{0.77}$, after which evaporation proceeds without change on passing from the solid to the vapor phase.

Chromium carbides evaporate in stages: $Cr_{23}C_6$ becomes impoverished in chromium and is converted into Cr_7C_3, which in turn is converted into Cr_3C_2 with evolution of chromium. Analysis of the available research data on the behavior of transition-metal carbides [336, 337, 429, 463, 673, 866] in vacuum at high temperatures has led to the conclusion that the carbides of metals of subgroups IVa and Va dissociate in accordance with the equation

$$Me\,C_{x(\text{solid})} = Me_{(\text{gas})} + x\,C_{(\text{gas})}\,.$$

The heats of formation of carbides decrease on passing from carbides of metals of subgroup IVa to those of the subgroups Va and VIa; this indicates a reduction in the Me — C bond energy due to the collective electrons and a strengthening of the Me — Me and C — C (covalent) bonds. According to [14], such a weakening of the Me — C bond energy is associated with an increase in the statistical weight of the stable d^5-configurations on passing from carbides of metals of subgroup IVa to those of subgroups Va and VIa, there being less likelihood of a breakdown of the stable sp^3-configurations of carbon and a greater separation of the sublattices of the metal and carbon in the carbides.

The results of researches to determine the enthalpy of the carbides of titanium, zirconium, and niobium of various compositions in the homogeneity range at 1300-1500°K have been analyzed in [949]. It is shown there that the experimental values of the specific heat of titanium and zirconium carbides at relatively low temperatures are lower than the additive values, the discrepancy diminishing with rise in temperature. The discrepancy also decreases with reduction of the carbon content in the carbide. For niobium carbides of all compositions the specific heats obtained experimentally are 5% higher than the additive values at relatively low temperatures. The authors of [949] associate this change in the specific heat with the total strength of the chemical bond in the compound: the stronger the bond, the lower should be the specific heat.

TABLE 2. Thermodynamic and Thermo-

Phase	Melting point, °C	Heat of formation from elements, ΔH_{298}		Change in free energy of formation from elements, ΔF_{298}		Entropy
		kJ/mole	kcal/mole	$\frac{kJ}{(deg \cdot mole)}$	$\frac{kcal}{(deg \cdot mole)}$	$\frac{J}{(deg \cdot mole)}$
1	2	3	4	5	6	7
Li_2C_2	—	+59.4	+14.2	—	—	—
Na_2C_2	800 (decomp.)	—17.18	—4.1	—35.558	—8.500	—
Be_2C	2400	91.3±±21	21.8±5	88.8±21	21.2±5	—
BeC_2	—	240.5	57.4	—	—	—
Mg_2C_3	—	79.6±±33.5	19±8.0	—75.36	—18.00	92.1
MgC_2	—	88±21	21±5.0	—87.92	—21.0	58.7±±10.5
CaC_2	2300	59.1±±8.4	14.1±±2.0	64.306	15.36	70.4±±2.1
SrC_2	1900 (decomp.)	84.6	20.2	—	—	—
BaC_2	1770—2300 (decomp.)	81.70	19.5	—	—	14.2
ScC	1800	—	—	—	—	—
YC	1950	—	—	—	—	—
Y_2C_3	1800	—	—	—	—	—
YC_2	2190	—	—	—	—	—
La_2C_3	1425	—	—	—	—	—
LaC_2	2360	159.1	38.0	—	—	—
Ce_2C_3	1535	—	—	—	—	—
CeC_2	2290	—	—	—	—	—
Pr_2C_3	1560	—	—	—	—	—
PrC_2	2120	—	—	—	—	—
NdC_2	2207	—	—	—	—	—

physical Properties of Carbides

S_{298}	Specific heat $C_{p_{298}}$		Thermal conductivity λ_{298}		Coeff. of thermal expansion α_{298}, 10^{-6} deg^{-1}	References
cal/(deg · mole)	J/(deg · mole)	cal/(deg · mole)	W/(m · deg)	cal/(cm·sec·deg)		
8	9	10	11	12	13	14
—	—	—	—	—	—	[65]
—	—	—	—	—	—	[65, 900]
—	49.0	11.69	43	0.102	7.47	[88, 92]
—	—	—	—	—	—	[65]
22.0	—	—	—	—	—	[65, 103]
14.0±±2.5	—	—	—	—	—	[65, 103]
16.8±±0.5	61.42	14.66	—	—	—	[60, 65, 103]
—	—	—	—	—	—	[152, 933]
3.4	—	—	—	—	—	[150, 152, 933]
—	—	—	—	—	11.4	[20]
—	—	—	—	—	1.36	[20]
—	—	—	—	—	—	[20]
—	—	—	—	—	11.1	[20]
—	—	—	—	—	—	[20]
—	—	—	—	—	13.7	[65, 20]
—	—	—	—	—	—	[20]
—	—	—	—	—	10.8	[20]
—	—	—	—	—	—	[20]
—	—	—	—	—	11.2	[20]
—	—	—	—	—	14.5	[20]

TABLE 2

Phase	Melting point, °C	Heat of formation from elements, ΔH_{298}		Change in free energy of formation from elements, ΔF_{298}		Entropy
		kJ/mole	kcal/mole	$\frac{kJ}{(deg \cdot mole)}$	$\frac{kcal}{(deg \cdot mole)}$	$\frac{J}{(deg \cdot mole)}$
1	2	3	4	5	6	7
ThC_{1-x} ($x \approx 0$)	2625	29.3± ±25.1	7±6	26.81	6.4	50.23
ThC_{2-x} ($x \approx 0$)	2655	192.7± ±25.1	46±6	136.07	32.5	80.4
UC_{1-x} ($x \approx 0$)	2315	90.9± ±4.2	21.7± ±1.0	—	—	59.9± ±4.2
U_2C_3	1820 (decomp.)	205.3± ±18	49±4.0	—	—	116.9± ±12.6
UC_{2-x} ($x \approx 0$)	2350	96.4± ±8.4	23.0± ±2.0	—	—	67.9± ±2.1
PuC_{1-x} ($x = 0.15$)	1850 (decomp.)	15.5± ±13	3.7±3.1	—	—	—
Pu_2C_3	1900	7.1	1.7	—	—	—
TiC_{1-x} ($x \approx 0$)	3257± ±150	183.7	43.85	236.8	56.5	24.3± ±0.4
ZrC_{1-x} ($x \approx 0$)	3530	201.1 ±6.28	48±1.5	181.72	43.4	35.6± ±6.3
HfC_{1-x} ($x \approx 0$)	3890± ±150	228.1 ± 4.2	54.2 ±1.0	199.0	47.53	28.1± ±8.2
V_2C_{1-x} ($x \approx 0$)	—	147.5	35.2	77.04	18.40	—
VC_{1-x} ($x = 0.12$)	2830	101.0	24.1	—	—	28.37± ±0.4
Nb_2C_{1-x} ($x \approx 0.14$)	—	78.3	18.7	82.2	20.59	30.4
NbC_{1-x} ($x = 0$)	3760	134.9	32.2	136.66	32.64	37.26
Ta_2C_{1-x} ($x \approx 0$)	3400	71.2	17.0	140.68	33.6	83.3
TaC_{1-x} ($x = 0.018$)	3880	140.9	33.7	159.52	38.0	42.32

(continued)

S_{298}	Specific heat $C_{p_{298}}$		Thermal conductivity λ_{298}		Coeff. of thermal expansion $\alpha_{298}\cdot 10^{-6}\ deg^{-1}$	References
cal/(deg · mole)	J/(deg · mole)	cal/(deg · mole)	W/(m · deg)	cal/(cm·sec·deg)		
8	9	10	11	12	13	14
12.0	—	—	28.9	0,069	6,53	[65, 197, 209—211]
19.2	—	—	23.9	0.057	8,5	[65, 197, 209—211, 213—215]
14.3±0.1	50,7	12,1	23,9	0.057	12.4	[199, 251, 256]
27.9±3.0	89.4	26.0	—	—	—	[251]
16.2±0.5	60.8	14.5	34.3	0.082	—	[231, 251]
—	—	—	—	—	—	[211]
—	—	—	—	—	—	
5,8±0.1	50,57	12.07	11	0.0263 (20—100 °C)	8.31±±0.68	[317, 368, 338, 339, 350, 838]
3,5±1.5	—	—	20.5	0.049	6.95±±0.25	[65, 152, 368, 397]
6.7±2 $HfC_{0.45}O_{0.05}$	54.7	13,05	6.3	0.015	6.59 ±0.04	[65, 152, 399, 403, 759]
—	—	—	—	—	—	[599]
6.77±±0.1	42.6	10.17	24.7	0.059	7.2±0.6	[25, 173, 362 397, 412, 424, 599]
7.25	—	25	—	—	7.0±0.3	[152, 457, 600, 873]
8.9	—	8.92	14.2	0,034	6.5	[65, 152, 457, 459, 873]
19.9	—	—	—	—	—	[152, 467]
10.11	32.5	7.77	22.2	0.053	8.29	[129, 152, 368, 467, 873]

TABLE 2

Phase	Melting point, °C	Heat of formation from elements, ΔH_{298} kJ/mole	Heat of formation from elements, ΔH_{298} kcal/mole	Change in free energy of formation from elements, ΔF_{298} $\frac{kJ}{(deg \cdot mole)}$	Change in free energy of formation from elements, ΔF_{298} $\frac{kcal}{(deg \cdot mole)}$	Entropy $\frac{J}{(deg \cdot mole)}$
1	2	3	4	5	6	7
$Cr_{23}C_6$	1500	68.6	16.38	70.14	16.74	106.0
Cr_7C_3	1780	178.15	42.51	183.35	43.76	201.1
Cr_3C_2	1895	88.0	21.01	88.83	21.20	85.5
Mo_2C_{1-x} $(x \approx 0)$	2690	—17.6	—4.2	11.73	2.8	80.0
W_2C_{1-x} $(x \approx 0)$	2730	46.05	11.0	48.98	11.70	—
WC	2870	35.2± ±12.6	8.4± ±3.0	37.60	8.98	35.58
$Mn_{23}C_6$	—	13.8	3.3	—	—	—
Mn_3C	—	15.1	3.6	14.27	3.41	98.9
Fe_3C	1650	22.6	—5.4	—19.05	—4.55	101.4
Co_3C	2300	—33.9	—9.3	—37.7	—9.0	95.7
Co_2C	—	—16.8	—4.0	—19.13	—4.57	74.5
Ni_3C	2200	—38.5	—9.2	—31.83	—7.6	106.4
B_4C_{1-x} $(x \approx 0)$	2350	57.8	13.8	50.66	12.10	27.1
Al_4C_3	2100	195.7± ±41.9	46.7± ±10	150.0	35.8	131.1
SiC	2827 (decomp.)	66.2	15.8	51.39	12.27	16.55

(continued)

S_{298}	Specific heat $C_{p_{298}}$		Thermal conductivity λ_{298}		Coeff. of thermal expansion α_{298}, 10^{-6} deg^{-1}	References
cal/(deg · mole)	J/(deg · mole)	cal/(deg · mole)	W/(m · deg)	cal/(cm·sec·deg)		
8	9	10	11	12	13	14
25.3	108.5	25.89	—	—	—	[368, 483, 492, 497]
48.0	209.16	49.92	—	—	9.4	
20.42	98.00	23.38	—	—	117	
19.1	—	—	6.7	0.076	7.8±±0.5	[60, 103, 152, 553]
—	—	—	29.0	0.07	1.2	[65, 553]
8.5	35.7	8.53	29.0	0.07	5.2	[103, 553, 900]
—	—	—	—	—	—	[587, 589, 596]
23.6	—	—	—	—	—	[65, 589, 593, 594]
24.2	68.4	25.5	—	—	—	[86, 152, 664, 900]
22.9	—	—	—	—	—	[60]
17.8	—	—	—	—	—	[65, 103]
25.4	—	—	—	—	—	[152, 368]
6.47	—	12.55	—	—	4.5	[152, 684, 697]
31.3	—	31.4	—	—	—	[103, 152, 174, 900]
3.95	0.955	0.228	8.0	0.02	5—7	[103, 152, 742]

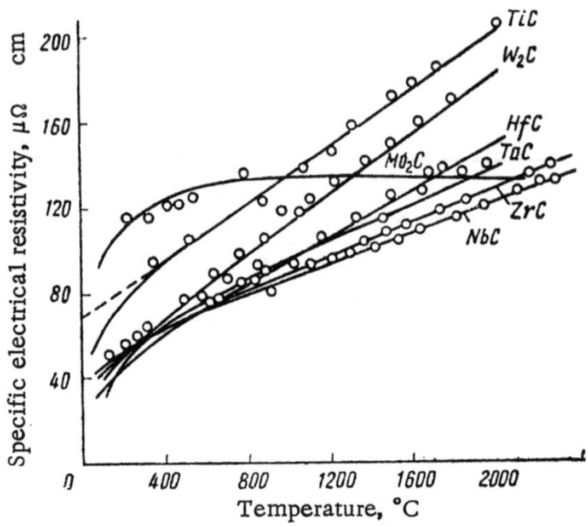

Fig. 12. Temperature dependence of the specific electrical resistivity of carbides of metals of subgroups IVa and Va.

The boiling points of the carbides are very high, ranging at atmospheric pressure from 2537°C for Be_2C to 6000°C for WC [152].

The coefficients of thermal expansion of transition-metal carbides are of the same order as those for the metals, diminishing with increase in the atomic number of the element in the group.

The thermodynamic and thermophysical properties of the carbides are given in Table 2.

Electrophysical and Magnetic Properties

As regards their electrical properties, carbides may be divided into two groups: metal carbides, having metallic conductivity, and carbides of nonmetals, having a conductivity typical of semiconductors. The transition-metal carbides belong mainly to the first group.

The temperature dependence of the specific electrical resistivity of transition-metal carbides is shown in Fig. 12 [903].

The positive thermal coefficients of the specific resistivity, like the absolute resistivity values, indicate the metallic nature of the conductivity in the carbides.

According to [397], the specific electrical resistivity of titanium and zirconium carbides obtained by deposition from the gaseous phase, i.e., having a preferred orientation and being little different from single crystals, is higher than that of the same compounds produced by powder metallurgy. Thus, ρ is 200 $\mu\Omega \cdot$ cm for $TiC_{0.92}$ and 182 $\mu\Omega \cdot$ cm for $ZrC_{0.964}N_{0.002}$.

By systematic investigations of the specific electrical resistivity and the coefficients of the thermal emf and Hall effect for carbides of the transition metals of groups IV-VI, and by an analysis of the results, it has been possible to estimate the magnitude of the contributions made by the electron and hole components to the conductivity of carbides [15, 16, 19].

In the case of charge carriers of the two signs

$$R = \frac{1}{e} \cdot \frac{n_- u_-^2 - n_+ u_+^2}{(n_- u_- + n_+ u_+)^2},$$

$$\rho = \frac{1}{e} \cdot \frac{1}{n_- u_- + n_+ u_+},$$

where R is the Hall coefficient, ρ is the specific resistivity, n_- and n_+ are the concentrations, and u_- and u_+ are the mobilities of electrons and holes, respectively, the quantity

$$\delta = \frac{R}{e\rho^2} = n_- u_-^2 - n_+ u_+^2,$$

which characterizes the relative contributions of electrons and holes to the electrical conductivity, has been calculated in [19]. A positive value of δ indicates predominantly electron conductivity and a negative value predominantly hole conductivity. For carbides, the value of δ is positive, decreasing on passing from carbides of metals of subgroup IVa to those of subgroups Va and VIa; this indicates a predominantly electron conductivity, the proportion decreasing in the series $Me_{IV}C - Me_{V}C - Me_{VI}C$ and the proportion of hole conductivity increasing correspondingly. On passing from titanium carbide to zirconium and hafnium carbides and

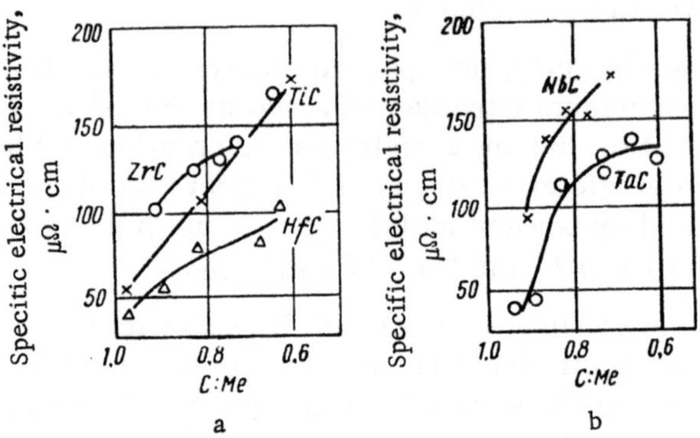

Fig. 13. Dependence of the electrical resistivity of carbides on carbon content in the homogeneity range.

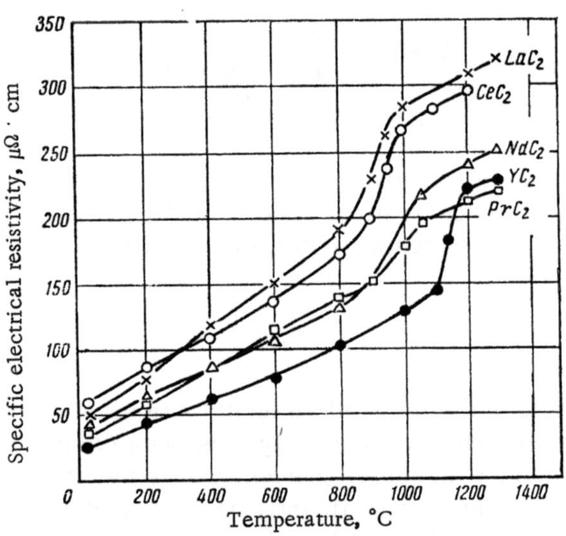

Fig. 14. Temperature dependence of the electrical resistivity of dicarbides of the rare-earth metals.

from vanadium carbide to niobium and tantalum carbides, the absolute value of δ increases, and, consequently, the proportion of electron conductivity also increases.

Investigations into the electrophysical properties of monocarbides of metals of subgroups IVa and Va in the homogeneity range have shown [13, 347, 391, 703, 715, 948] that the Hall and thermal emf coefficients increase with the carbon content (Fig. 13a, b). The negative values of the Hall and thermal-emf coefficients for TiC_{1-x} indicate the predominantly electron nature of the conductivity over the whole homogeneity range, the proportion of electron conductivity increasing with the approach to the stoichiometric composition. According to data in [703, 715], the reduction in resistivity with rise in carbon content is due to insufficiently high purity of the carbides investigated. Carbides of higher purity showed a linear relationship.

The higher carbides of the rare-earth metals (MeC_2) also have a metallic, predominantly electron, type of conductivity. With rise in temperature to 800°C, the resistivity increases almost linearly; in the range 900–1370°C there is a sharp discontinuity in the resistivity values, after which the relationship again becomes linear (Fig. 14). A similar anomaly is observed in the temperature dependence of the thermal-emf coefficient and the thermal expansion [924]. This suggests that a phase transformation takes place in the carbides of the rare-earth metals.

From a study of the electrophysical properties of dicarbides [20], it has been possible to calculate the current-carrier concentrations, which are similar for LaC_2, PrC_2, and NbC_2, being 0.5 electrons per Me atom. Since these carbides crystallize in a tetragonal lattice containing four metal atoms and four C_2 groups per unit cell, it has been assumed that three C_2 groups exist in the form $(C \equiv C)^{2-}$ and one in the form $(C = C)^{4-}$. Consequently, of the 12 valence electrons in the metal atoms, 10 participate in the Me–C bond and two act as conduction electrons. This suggested model of the structure of dicarbides agrees well with data from the analysis of the gaseous products of decomposition with water, which show that the principal products are acetylene, ethylene, and hydrogen. The model also accounts for the high conductivity of the dicarbides.

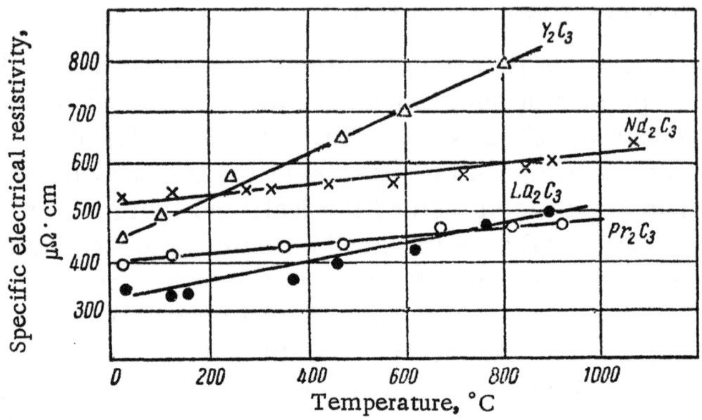

Fig. 15. The temperature dependence of the electrical resistivity of the sesquicarbides of the rare-earth metals.

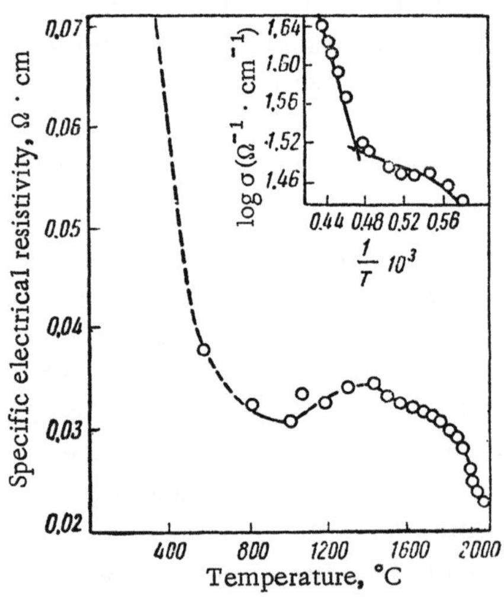

Fig. 16. The temperature dependence of the electrical resistivity of commercial boron carbide.

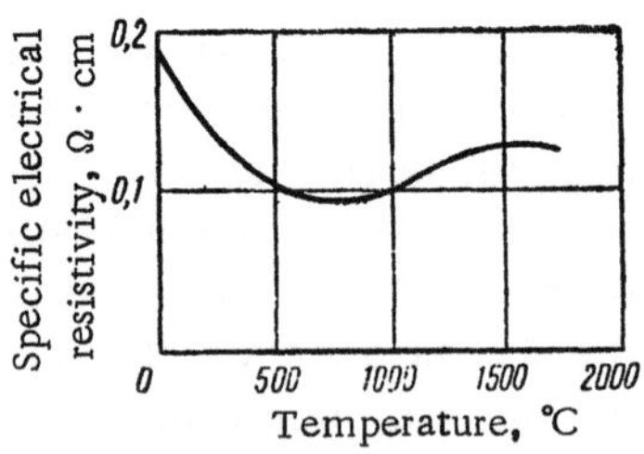

Fig. 17. The temperature dependence of the electrical resistivity of silicon carbide.

The specific electrical resistivity of the sesquicarbides of the rare-earth metals is an order of magnitude greater than that of the dicarbides, and it increases linearly with rise in temperature (Fig. 15).

Boron and silicon carbides are distinguished by the semiconducting nature of their conductivity. Figure 16 shows the temperature dependence of the specific electrical resistivity of commercial boron carbide; it shows that the resistivity first decreases as a result of excitation of the impurity electrons, after which it increases somewhat owing to scattering at the thermal vibrations of the lattice and then decreases again at the transition to intrinsic conductivity of the carbide [909]. Silicon carbide is a typical impurity semiconductor, and its resistivity depends to a considerable extent on the impurities and lattice defects, including the presence of other elements and departure from the stoichiometric composition. The electrical conductivity of commercial silicon carbide varies between 10 and $10^{-12}\ \Omega^{-1} \cdot cm^{-1}$ [711]. The character of the temperature dependence of the resistivity of silicon carbide (Fig. 17) is the same as for boron carbide.

An investigation into the conductivity of SiC single crystals obtained by deposition from a gaseous mixture of $SiCl_4$ and toluene, has led to the conclusion that in p-type SiC the holes in the main filled band and the electrons in the impurity conduction band act as current carriers; in n-type SiC the electrons in the main conduction band and the electrons in the impurity band act as the current carriers [711]. It has been established in [702] that the impurities nitrogen and phosphorus impart n-type conductivity to SiC single crystals, while boron and aluminum impart p-type conductivity. In the first case nitrogen gives a green color to the crystals and in the second case aluminum gives a blue or black color.

Magnetic Properties. The magnetic properties of carbides have been studied mainly in carbides of certain transition metals of groups IV and VI and rare-earth metals. Since the transition-metal carbides are chiefly interstitial phases in which all the outer electrons are paired, they are weakly paramagnetic. Calculated magnetic moments of dicarbides of the rare-earth elements are somewhat less than those of trivalent ions (except YbC_2), but similar to them (Fig. 18); this indicates little breakdown in the $4f$-states of the rare-earth metals on the formation of the carbides [904].

TABLE 3. Electrical and Magnetic Properties of Carbides

Phase	Specific electrical resistivity, $\mu\Omega \cdot cm$	Thermal coeff. of elec. resistiv., $\alpha \cdot 10^{-3}\ deg^{-1}$	Coeff. of thermo-emf, $\mu V/deg$	Hall constant, $R \cdot 10^4$ cm^3/C	Work function		Richardson constant, $A/cm^2 \cdot deg^2$	Trans. temp. to super-conducting state, °K	References
					10^{19} J	eV			
Be_2C	$1.1 \cdot 10^6$	—	—	—	—	—	—	—	[88]
ScC	274	—	—	—	—	—	—	—	[25]
YC	$4.54 \cdot 10^4$	—	—34.6	—	—	—	—	—	
Y_2C_3	450	+0.98	—6.4	—	—	—	—	—	
YC_2	30	+3.63	—12.0	—11.95	—	—	—	—	
La_2C_3	334	+0.56	—	—4.41	—	—	—	—	[20]
LaC_2	45	+4.10	—9.0	—5.03	—	—	—	—	
Ce_2C_3	383	—	—	—	—	—	—	—	
CeC_2	60	+2.18	—2.4	—2.80	—	—	—	<1.28	[152, 20]
Pr_2C_3	396	±0.24	—	—	—	—	—	—	[20]
PrC_2	39	+3.85	—9.1	—5.38	—	—	—	—	[152, 20]
Nd_2C_3	500	+0.26	—	—	—	—	—	—	[20]
NdC_2	44	+2.65	—10 0	—4.38	—	—	—	—	
ThC_{1-x} ($x \approx 0$)	25	—	—	—	—	—	—	9±1	[209, 213]
ThC_{2-x} ($x \approx 0$)	30	—	—	—	5.6	3.5	550	—	[18, 209]
UC_{1-x} ($x \approx 0$)	47	—	—	—	7.15	4.47	$4.3 \cdot 10^5$	<1.20	[152, 213, 264]
PuC_{1-x} ($x = 0.15$)	230	—	—	—	—	—	—	—	[199]

TiC_{1-x} (x=0.012)	52.5	+1.[illegible]6	—12.5	—6.7	5.36	3.35	2.5	1.15	[19, 152, 370, 855, 864]
ZrC_{1-x} ($x\approx0$)	68	+0.95	—11.3	—9.42	3.49	2.18	134	4.1	[152, 264, 855, 864]
HfC_{1-x} ($x\approx0$)	45	+0.95	—11.1	—12.4	3.26	2.04	$\approx10^{-5}$	<1.23	[19, 152, 405, 411, 413, 864]
VC_{1-x} (x=0.12)	98	—	—	—0.9	6.1	3.85	—	<1.20	[152, 855]
NbC_{1-x} ($x\approx0.15$)	51.1	+0.86	—4.0	—1.32	3.57	2.23	$\sim10^{-6}$	14.0	[18, 19, 413, 901]
TaC_{1-x} (x=0.04)	42.1	+1.07	—5.0	—1.1	5.02	3.14	0.30	11.0	[18, 19, 152, 370, 901]
$Cr_{23}C_6$	127±2	+1.72±0.11	—2.76±0.02	—1.2±0.2	—	—	—	<1.20	[152, 518]
Cr_7C_3	109±4	+1.06±0.05	—7.1±0.03	—0.38±0.03	—	—	—	<1.20	[152, 518]
Cr_3C_2	75±5	+2.33±0.04	—6.7±0.5	—0.47±0.03	—	—	—	<1.20	[152, 518, 520]
Mo_2C	71	—3.78	—1.9	—	—	—	—	2.78	[19, 152, 901]
W_2C	75.7±0.1	1.95	—8.17	—	—	4.58	190	2.74	[19, 152, 901]
WC	19.2±0.3	0.495	—23.3	—	—	—	—	<1.28	[19, 152]
$B_{6.5}C$	12	—	+180	—	—	—	—	—	[691]
B_4C_{1-x} ($x\approx0$)	10^6	+0.032	+80	—	—	—	—	<1.28	[152, 690]

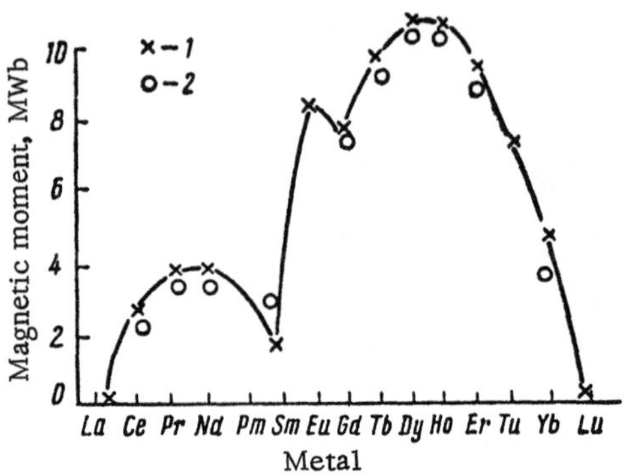

Fig. 18. Magnetic moments of the rare-earth metals and their carbides. 1) Me; 2) MeC_2.

Superconductivity. The highest transition temperatures to the superconducting state are observed in the carbides of niobium and molybdenum; the temperatures are higher in phases containing more carbon than in those containing less. Thus, whereas for NbC, T_c is 14°K, for Nb_2C it is only 1.98°K. The values of T_c for carbides in the homogeneity range fall with decrease in the carbon content, and at small contents the superconductivity practically disappears.

Thermionic-Emission Properties. From a consideration of the thermionic-emission properties it may be concluded that the work function for transition-metal carbides is lower than that of the respective pure metals. Thus, for titanium it is 4.09 eV and for titanium carbide, 3.35 eV; for zirconium it is 3.9 eV, and for zirconium carbide, 2.18 eV, etc. [910].

Maximum values of the work function are observed in the carbides of titanium, vanadium, and molybdenum.

The electrophysical and magnetic properties of the carbides are given in Table 3.

Physicomechanical Properties

As can be seen from the data in Table 4, the transition-metal carbides, carbide phases of the rare-earth metals of high carbon content, and nonmetal carbides have a high microhardness.

TABLE 4. Mechanical Properties of Carbides

Phase	Microhardness		Modulus of elasticity		References
	MN/m^2	kgf/mm^2	MN/m^2	kgf/mm^2	
Be_2C	26270	2690	320000	32000	[88, 152]
ScC	26650	2720	—	—	[180]
ThC_{1-x} ($x\approx 0$)	8330	850	—	—	[209]
ThC_{2-x} ($x\approx 0$)	5880	600	—	—	
UC_{1-x} ($x\approx 0$)	—	—	220000	22050	[231]
U_2C_3	—	—	—	—	[220]
UC_{2-x} ($x\approx 0$)	—	—	—	—	[231]
TiC_{1-x} ($x\approx 0$)	27930±39,2	2850±40	315000	32000	[277, 350, 367, 369, 370]
ZrC_{1-x} ($x\approx 0$)	28660±1800	2925±184	350000	35500	[152]
HfC_{1-x} ($x\approx 0$)	28540	2913	350000	35900	[152, 405]
VC_{1-x} ($x=0,15$)	20520	2094	270000	27600	[152, 424]
Nb_2C_{1-x} ($x\approx 0$)	20800±1950	2123±199	—	—	[152]
NbC_{1-x} ($x=0,11$)	19220±940	1961±96	—	—	
Ta_2C_{1-x} ($x\approx 0$)	16800±1560	1714±159	—	—	
TaC_{1-x} ($x=0,06$)	15670±480	1599±49	—	—	[152, 432]
$Cr_{23}C_6$	16500	1650	—	—	[803]
Cr_7C_3	21550	2200	—	—	[803]
Cr_3C_2	22400	2280	370000	38000	[152, 519]
Mo_2C_{1-x} ($x\approx 0$)	14490±2350	1479±24	220000	22100	[103, 152, 908]
W_2C_{1-x} ($x\approx 0$)	29400	3000	420000	42800	[152]
WC	9040	923	700000	72200	[152, 908]
$B_{6,5}C$	54320—56260	5600—5800	—	—	[680]
B_4C_{1-x} ($x\approx 0$)	48450	4950	—	—	[152]
SiC	32400	3340	380000	39400	

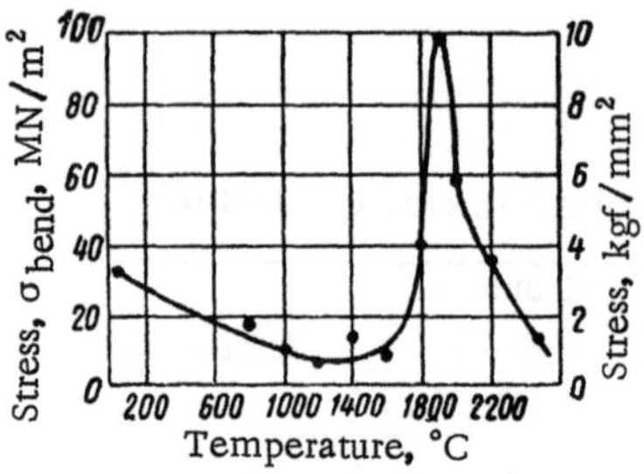

Fig. 19. Temperature dependence of the bending strength of titanium carbide.

The microhardness decreases on passing from the carbides of metals of subgroup IVa to those of subgroups Va and VIa. Data on the microhardness of the carbides of metals of groups VII and VIII are still meager, but their hardness is known to be considerably lower than that of carbides of metals of the subgroups IVa, Va, and VIa. As was shown in [905, 906], the hardness of refractory metallic compounds is connected with the energy of atomic interaction, i.e., with the energy of the crystal lattice, which in turn is determined by the electron structure of the elements forming the compound. It was shown in [836] that the microhardness of refractory compounds, in particular carbides, is determined by the statistical weight of the stable electron configurations of the metal and carbon atoms and increases with rise in the statistical weight of the sp^3-configurations of carbon.

Microhardness studies in the homogeneity range of carbides [374, 381, 473] have established that the microhardness of zirconium and tantalum carbides falls linearly with decrease in carbon content, and on extrapolation to zero carbon content the microhardness values correspond to those of the pure metals in the case of carbides of metals of subgroup IVa. The highest microhardness values occur in the carbides of boron and silicon, which are characterized by strong directional covalent bonds between the atoms of carbon and boron or silicon.

The strength properties of carbides have not yet been adequately studied. An investigation into the short-time bending strength of titanium carbide specimens with a porosity of 20-25% [908] has shown (Fig. 19) that the strength varies very little with rise in temperature up to 1600°C. With further rise in temperature the strength increases rapidly, reaching 104 MN/m² (10.4 kgf/mm²) at 1900°C; beyond this any further rise in temperature leads to a sharp fall in the strength. Similar changes in strength also occur in specimens of other transition-metal carbides. For such carbides the maximum strength in bending occurs at $0.6T_{mp}$.

Chemical Properties

Carbides can be divided into two groups, depending on whether or not they are decomposed by reaction with water.

Among those that are decomposed by water are the carbides of metals of groups I and II of the periodic system, aluminum, and also the carbides of the rare-earth metals and the actinides. The carbides of alkali and alkaline-earth metals are products of the replacement of the hydrogen atoms in acetylene by metals — acetylides — and decompose with evolution of acetylene on reaction with water and dilute mineral acids. The behavior is somewhat different in the case of the carbides of beryllium (Be_2C) and aluminum, which on reaction with water decompose with the evolution of methane and which consequently are products of the replacement of the hydrogen atoms in methane by metals, and also in the case of magnesium carbide, which decomposes on reaction with water to evolve methyl acetylene $CH_3-C\equiv C-H$. On reaction with water, carbides of the rare-earth metals and actinides decompose with the evolution of a mixture of hydrocarbons. According to mass-spectrographic studies [7], on decomposition by water of the carbides of the rare-earth metals Me_3C_{1-x}, the gaseous products consist of methane and hydrogen.

The decomposition products of the Me_3C_2 and MeC_2 phases are acetylene and hydrogen; the quantity of hydrogen liberated on the decomposition of La_2C_3 is greater than that on decomposition of LaC_2.

In the literature there are many papers dealing with the composition of the gaseous products of the decomposition of carbides of rare-earth metals and actinides by water [8, 64, 204, 209, 770, 785, 789, 791, 911]. The results given in these papers are often very divergent (Table 5). This is probably due to the compositions of the carbides investigated, which frequently are not even given or do not include data on the combined-carbon content.

Carbide phases of the rare-earth metals close to the stoichiometric composition were used in [8], in which decomposition was carried out by treating carbide samples with water in quartz reactors through which carbon dioxide was passed and the gaseous products evolved were afterwards analyzed by the chromatographic method [787]. The most reliable data on the composition of the

TABLE 5. Composition of the Gaseous Products of Decomposition of Dicarbides of the Rare-Earth Metals by Water

Phase	Content, %							References
	H_2	CH_4	C_2H_2	C_2H_4	C_2H_6	C_3H_8	C_3H_6 and its homologs	
YC_2	4.5	—	71.7	19.0	—	—	—	[64]
	5.03	15.6	58.6	10.98	8.04	—	—	[8]
LaC_2	1.03	27.2	71.75	—	—	—	—	[64]
	10.02	—	67.19	6.74	12.48	1.36	2	[789]
	15.6	Not detect.	53.0	11.2	20.2	—	—	[8]
CeC_2	—	20.27—21.48	75—75.5	3.52—4.13	—	—	—	[64, 791]
	13.4	—	68.0	5.91	8.81	0.96	5.55	[789]
	15.9	Not detect.	58.23	10.49	15.35	Not detect.	—	[8]
PrC_2	—	30.0	66.2	2.50	—	—	—	[64, 791]
	6.79	—	78.16	8.18	10.68	0.75	1.22	[789]
	7.92	Not detect.	63.92	11.90	16.26	Not detect.	—	[8]
NdC_2	—	27.4	66.2	6.34	—	—	—	[64, 791]
	12.61	—	67.82	7.61	8.41	1.79	1.28	[789]
	4.70	Not detect.	67.49	12.50	15.31	Not detect.	—	[8]
SmC_2	—	21.5	70.65	7.85	—	—	—	[64, 785, 791]
	5.1	—	71.1	7.9	12.2	1.28	2.32	[789]

GdC_2	17.60	Not detect.	52.8	14.8	13.3	Not detect.	—	[8]
$ThC_{0.96}O_{0.09}$	9.3	86.4	—	0.75	—	1.2	2.35—C_2H_3 1.4—C_3H_4	[209]
	17.1	29.4	47.7	5.8	—	—	—	[204]
ThC_{2-x}	59.6	3.1	15.0	2.8	10.7	Liquid and solid hydrocarbons	8.8	[770]
	16.6	33.2	49.8	—	—	—	—	[785]
$ThC_{1.84}O_{0.06}$	27.2	2.35	9.5	2.45	—	1.5	29.1—C_2H_3 1.5—C_3H_4 8.6—C_4H_6 11 3—C_4H_8 3.1—C_4H_{10} Higher hydrocarbons 2.0	[209]

TABLE 6. The Resistance of Carbides of Metals

Solvent	Insoluble			
	TiC_{1-x}	ZrC_{1-x}	HfC_{1-x}	NbC_{1-x}
HCl (density 1.19)	99/100	98/98	100/100	100/96
HCl (1:1)	100/97	98/95	96/43	100/99
H_2SO_4 (density 1.84)	100/88	97/0	100/0	100/0
H_2SO_4 (1:4)	100/97	98/76	100/88	100/98
HNO_3 (1,43)	0/0	83/5	60/0	100/(hydrolysis)
HNO_3 (1:1)	0/0	76/6	75/0	—
H_3PO_4 (1,7)	99/98	98/—	97/0	100/(hydrolysis)
H_3PO_4 (1:3)	98/99	96/88	98/90	90/99
$HClO_4$	100/0	97/2	—	—
$HClO_4$ (1:3)	100/0	99/84	—	—
$H_2C_2O_4$ (sat. solution)	100/100	98/92	97/90	99/99
$HCl+HNO_3$ (3:1)	0/0	14/6	0/0	92/(hydrolysis)
$H_2SO_4+HNO_3$ (1:1)	—	2/0	0/0	100/22
HNO_3+HF (4:1)	0	0	0	0
$H_2SO_4+H_3PO_4$ (1:1)	98/—	97/—	99/0	91/0
$H_2SO_4+H_2C_2O_4$ (1:1)	99/34	96/0	99/95	100/95
$H_2SO_4+H_2C_2O_4$ (1:1)	99/99	96/91	98/66	—
$H_2SO_4+H_3PO_4$ (1:1)	—	—	—	—/99
NaOH:				
20% solution	100/100	100/100	100/98	99/100
10% solution	100/100	100/100	100/98	99/—
20% + bromine water (4:1)	63/86	98/87	88/64	100/84
20% + H_2O_2 (4:1)	—	53/3	34/0	71/88
20% + $K_3[Fe(CN)_6]$ (10% solution)	—	82/37	86/24	—

*Powders with a particle size of 40-50 μm were used. The numerator gives the insoluble residue obtained by treating the carbides for 24 h at of the Subgroups IVa-VIa in Various Media

residue*						
TaC_{1-x}	Cr_3C_2	Cr_7C_3	$Cr_{23}C_6$	Mo_2C_{1-x}	WC	B_4C_{1-x}
100/98	99/—	92.3/—	53/—	89/89	97/48	98/98
100/98	99/96	99/3.5	—1.3	88/83	96/92	—/97.6
100/0	—/—	—	—	89/0	91/1	98/98
99/93	100/68 (H_2SO_4 1:1)	99/1.6 (H_2SO_4 1:1)	52/1.3 (H_2SO_4 1:1)	90/83	96/95	—
100/99	100/99	100/99	100/98	0/0	63/1	97/97
99/98	100/99	—	—	0/0	72/10	—/96.9
98/(hydrolysis)	100/98	100/96	100/—	92/—	91/93	—
—/98	—	—	—	92/76	96/90	—
—	—	—	—	73/0	98/40	98/98
—	—	—	—	89/58	98/93	—
97/98	100/98	100/96	100/98	89/90	95/95	—
99/98	98/91	95/94	100/97	0/0	28/3	—
91/96	—/84	—/91	—/97	1/0	92/42	—/91.2
0	Insoluble	Insoluble	Insoluble	0	0	—/90.8
Hydrolysis	—/94	—/4	—/1.4	90/0	96/0	—
97/97	—	—	—	80/73	95/70	—
93/98	—	—	—	—	94/95	—
100/99	—	—	—	92/88	96/93	—
99/99	100/99	—/96	0/96	90/94	98/98	—/99.1
99/100	—	—	—	88/90	97/98	99/99
100/(hydrolysis)	—/88	—/86	—/86	65/60	70/60	—/99.6
62/ (hydrolysis)	—	—	—	31/36	88/87	—/98
—	—/61,5	—/53	—	69/66	68/68	—

20-25°C; the denominator gives the insoluble residue obtained by treating the carbides for 2 h at the boiling points of the respective solvents.

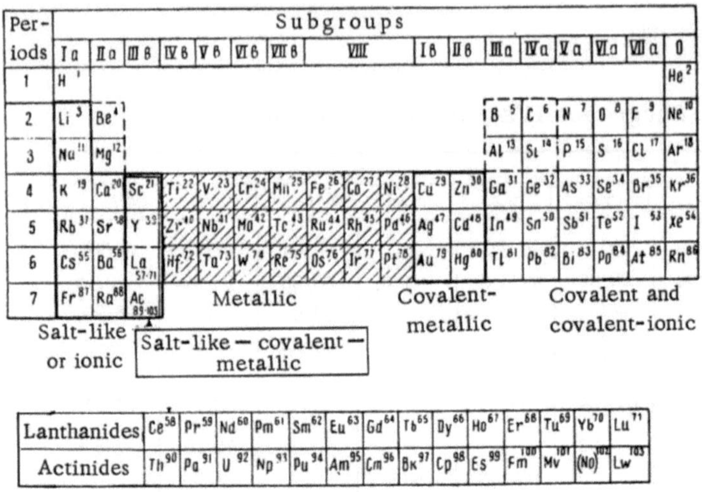

Fig. 20. Classification of carbides.

TABLE 7. Chemical Resistance of Carbides of Metals of Groups IVa and Va [894]

Medium	Insoluble residue, %					
	$TiC_{0.99}$	$ZrC_{0.98}$	$HfC_{1.00}$	VC_x	$NbC_{0.98}$	$TaC_{0.995}$
NH_4F, 5% solution	99.8	76.4	76.9	—	92.0	90.3
H_2O_2, 30%	10.0	97.2	99.6	5.3	0.0	83.7
$(NH_4)_2S_2O_8$, 25%	82.2	36.5	0.0	32.5	88.5	91.2
H_2O_2 + tartaric acid	3.3	95.6	96.5	64.5	0.0	57.0
H_2O_2 + citric acid	1.2	95.0	97.6	62.7	0,0	57.0
H_2O_2 + oxalic acid	0.0	Hydrolysis	0.0	59.2	0.0	—
H_2O_2 + saturated solution of complexone III	12.5	98.0	97.0	—	0.0	3.5
H_2O_2 + solution NH_4F	5.3	0.0	0.0	30.0	0.0	3.5
$(NH_4)_2S_2O_8$ + solution of complexone III	97.0	96.4	72.4	—	97.6	—
$(NH_4)_2S_2O_8$ + solution NH_4F	79.8	0.7	0.0	80.20	80.0	63.3
HCl + solution $(NH_4)_2S_2O_8$	50.2	44.6	0.0	60.0	97.2	98.4
HCl + H_2O_2	43.8	45.5	2.9	78.6	—	99.1
HCl + saturated solution of bromine water	15.6	8.5	1.1	45.6	87.5	97.7
H_2SO_4 (1 : 4) + solution $(NH_4)_2S_2O_8$ (25%)	70.3	4.3	0.0	15.6	64.0	97.7
H_2SO_4 (1 : 1) + solution $(NH_4)_2S_2O_8$ (25%)	45.3	2.2	0.0	—	21.8	—
H_2SO_4 (1:4) + H_2O_2 (30%)	0.25	0.75	0.0	15.4	13.2	—

gaseous products of decomposition of actinide carbides by water are those reported in [209].

Analysis of the data on the composition of the gaseous products of decomposition of dicarbides of the rare-earth metals by water leads to the conclusion that maximum evolution of hydrogen and minimum evolution of acetylene occur in lanthanum and gadolinium dicarbides, which are characterized by stable $4f^0$ and $4f^7$ states and the presence of electrons in the d-state.

According to Hund's rule [9], the number of possible terms increases, and the probability of $f \rightarrow$ d electron transfers decreases correspondingly, on passing from cerium to praseodymium, neodymium, etc., to gadolinium. The reduced likelihood of d-electrons participating in the Me — C bonds leads to a weakening of these bonds and to an increase in the amount of acetylene given off during decomposition. The evolution of acetylene and ethylene confirms the model of the structure of the rare-earth metal carbides containing three $(C \equiv C)^{2-}$ groups and one $(C = C)^{4-}$ group which was suggested on the basis of a study of their physical properties.

Among the carbides which are not decomposed by water are those of the transition metals, boron, and silicon.

Transition-metal carbides are characterized by high chemical resistance [755, 894] and are not decomposed by the majority of mineral acids, by mixtures of them, or by alkaline solutions (Tables 6 and 7). In a number of cases the chemical resistance of the carbides of metals of subgroup Va is greater than that of carbides of metals of subgroups IVa and VIa. This may be tentatively explained by the fact that in carbides of metals in subgroup Va, particularly niobium and tantalum, the metal and carbon atoms have a very high statistical weight of the stable electron configurations and a very low statistical weight of the unlocalized electrons, which are mainly responsible for the tendency of a compound to interact with reagents and to transfer electrons to them. Hence, carbides of metals of subgroup IVa, which are characterized by a high statistical weight of unlocalized electrons, usually have a lower chemical resistance than those of metals of subgroup Va. This is also observed in the carbides of certain metals of subgroup VIa, for example, molybdenum and tungsten carbides, in which the statistical weight of the stable d^5-configurations of the atoms in the metal decreases owing to the transfer of part of the localized electrons to the carbon atoms. At the same time there is an increased concentration of electrons in the free state and a greater possibility of their transfer to the components of the reagent, with decomposition of the carbide concerned. Chromium carbides have a high chemical resistance owing to the fact that in the groupings of chromium and carbon atoms the statistical weight of the stable electron configurations is high and the concentration of unlocalized electrons between these groupings is small. Hence, the probability of the transfer of these electrons to atoms of the reagent is considerably reduced, and chromium carbides are decomposed with difficulty. The above assumption requires experimental confirmation, and researches in this direction are being carried out.

The highest chemical resistance is to be found in the carbides of boron and silicon, which are not decomposed by mineral acids or mixtures of them, including the mixture of nitric and hydrofluoric acids which decomposes almost all the transition-metal carbides (except those of chromium) (Table 6). The high chemical resistance of boron and silicon carbides may be attributed to the strong covalent bonds between the atoms of carbon and boron or silicon and to the very low concentration of unlocalized electrons.

From an analysis of the results of an investigation into the wetting of carbides by liquid metals [937], Samsonov et al. have been able to show that metals of subgroups IIIb and IVb hardly wet refractory-metal carbides at all, whereas the transition metals wet them very well. The nature of the reaction of carbides of metals of subgroup IVa differs from that of carbides of metals of subgroups Va and VIa. Carbides of metals of subgroup IVa are considerably more resistant to molten metals than are the carbides of metals of subgroups Va and VIa.

From an analysis of the complex of properties of carbides and their crystal structures, Meerson [1] has suggested a classification of carbides which has been somewhat modified by Samsonov [2, 17]. On the basis of the nature of the chemical bond responsible for the particular features of the electron structure, the crystal structures, and the physical and chemical properties of carbides, Samsonov has proposed the following classification (see Fig. 20).

1. Salt-like or ionic carbides formed by alkali and alkaline-earth metals having valence s-electrons with the first ionization potential from 3 to 7 eV.

2. Metals of subgroups Ib and IIb, which, as a result of their high ionization potentials (7.5–11 eV) which make it difficult for the s-electrons to transfer to the bond with the carbon atoms, either do not form carbides at all or form unstable ones. As a result of this they can be put in the first group or into a separate group of covalent-metallic carbides.

3. Covalent (boron and silicon) and covalent-ionic (beryllium and magnesium) carbides formed by elements having valence sp-electrons.

Evidently the tendency of the electrons in beryllium to undergo the transition $1s^2 2s^2 \rightarrow 1s^2 2s2p$ makes beryllium carbide similar to the covalent carbides formed by elements with outer p-electrons. In this carbide strong covalent bonds are formed between the beryllium atoms. The changeover from covalent bonds between carbon atoms to covalent bonds between beryllium atoms is responsible for the formula Be_2C, in contrast with the formula MeC_2 for the alkaline-earth metals. The strong covalent bond between beryllium atoms while the covalent bond is retained between the carbon atoms accounts for the high degree of asymmetry of the electron-density distribution and imparts to beryllium carbide the properties of an ionic compound, i.e., it is a covalent-ionic compound with a predominantly covalent bond.

Magnesium carbide is similar in composition and structure to the carbides of the alkaline-earth metals, but in the nature of its bond it is closer to the covalent-metallic carbides. The electrons in the magnesium atom, like those in the beryllium atom, are able to undergo the $s \rightarrow p$ transition, but as the quantum number of its valence electrons is higher, the statistical weight of the sp^3-configurations is less and they are energetically less stable. On this account magnesium forms the carbide MgC_2 and not Mg_2C like beryllium; in this case, part of the carbon atoms form strong covalent bonds of the $C \equiv C$ type on account of the s-electrons in the magnesium atom, and part of them form weaker $C - C$ bonds on account of the formation of sp-configurations by the magnesium atoms. In consequence, on decomposition with water it is not methane which is given off, as in the decomposition of Be_2C, nor is it acetylene, as in the decomposition of the carbides of the alkaline-earth metals, but it is methyl acetylene $CH_3 - C \equiv C - H$.

4. Metallic carbides formed by the sd-transition metals. To this group belong carbides consisting of the interstitial phase and Fe_3C-type carbides having structures similar to the interstitial phases; their properties are the result of a regular and monotonic change on passing from carbides of metals of subgroup IVa to those of metals of subgroups Va and VIa and group VIII.

5. Salt-like covalent-metallic carbides formed by sd$f$-transition metals. These are the carbides of scandium, yttrium, the lanthanides, and the actinides.

Chapter II

Methods of Producing Carbides

The principal methods of preparing carbides can be divided into five groups: synthesis from the elements, reduction of the metal oxides by carbon, deposition from the gaseous phase, electrolysis of molten salts, and chemical precipitation. A classification of the methods of obtaining carbides is given in Fig. 21.

Synthesis from the elements, carried out by melting in an electric arc furnace, is historically the first method by which many carbides were obtained [21, 22]. Electric arc melting units are now used for this purpose.

The majority of metals react with carbon at temperatures below their melting point. Consequently, the most widely used method for producing carbides is by heating a mixture of the metal and carbon at temperatures below the melting point. This method is carried out in a reducing gas atmosphere (hydrogen, carbon monoxide, methane) or in vacuum, and it is widely used for producing molybdenum and tungsten carbides. Recently, use has been made of a method of producing carbides in which a mixture of metal and carbon powders is heated under pressure — the hot-compaction method — by which it is possible to obtain directly dense components of the carbides concerned. The method is used to obtain dense specimens of the carbides of uranium, tantalum, and a number of other metals.

The increasing demands now being made on carbides necessitate the development of new methods of producing high-purity carbides. One such method is by synthesis carried out in a liquid-metal bath [954]. In this method the requirements of the bath

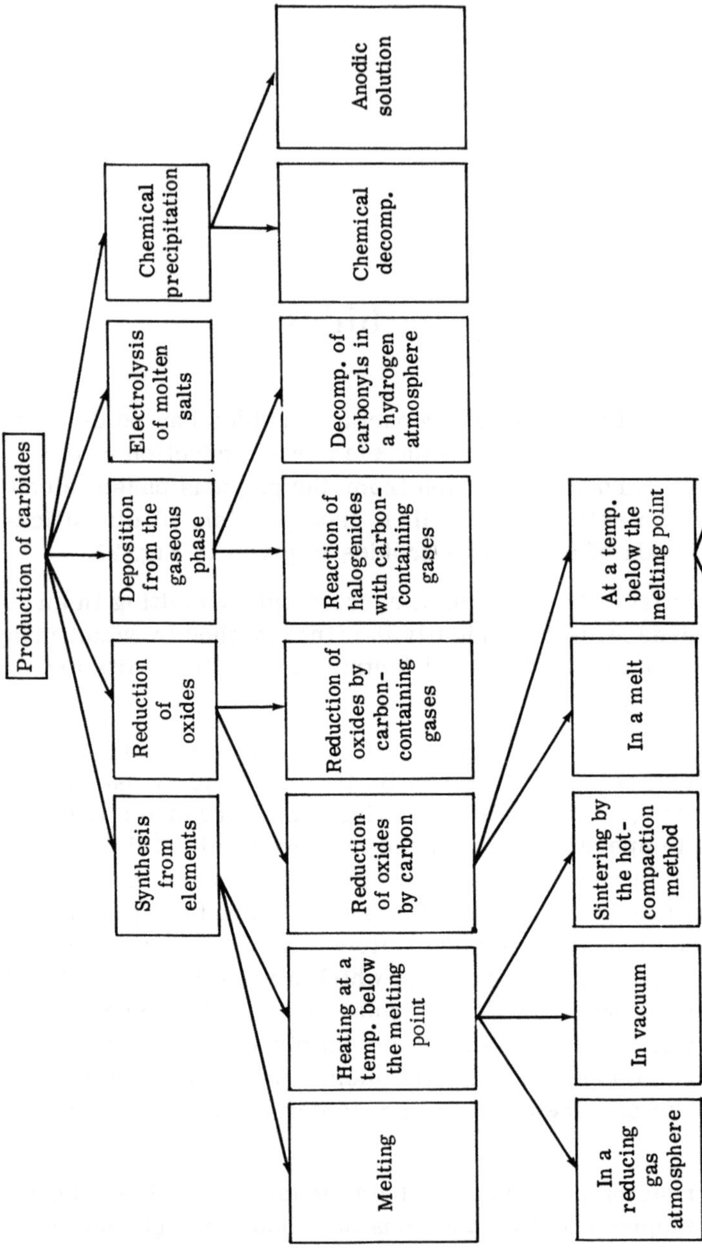

Fig. 21. Classification of methods of obtaining carbides.

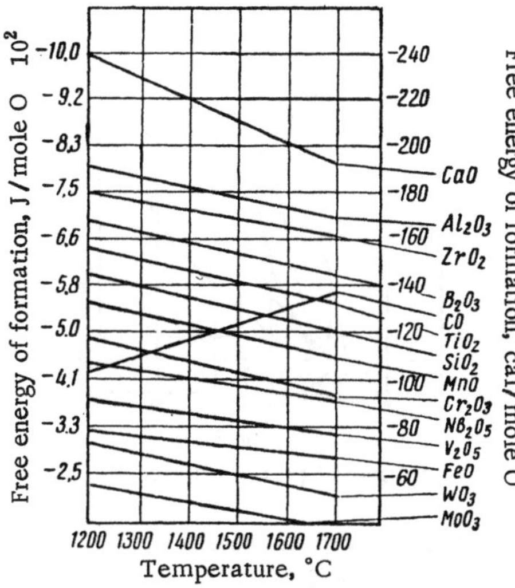

Fig. 22. Affinity of some transition metals for oxygen.

material are as follows:

1. Capacity to dissolve the transition metal forming the carbide.
2. Capacity to form unstable carbides or not to form them at all.
3. Capacity to be separated from the carbides by selective solution in chemically reactive media.

These requirements are satisfactorily met by metals of the iron group. It has been shown that by carrying out the synthesis in a liquid bath consisting of a 70:30 Fe — Ni alloy, and subsequently washing the reaction products in hydrochloric acid, it is possible to obtain transition-metal carbides close to the stoichiometric composition and containing less than 100 ppm oxygen, less than 10 ppm nitrogen, and less than 500 ppm other impurities.

Owing to the high cost of the powders of some metals, the most economical method of producing carbides is by reducing the oxides of the corresponding metals with carbon or carbon-containing gases. The oxides may also react with carbon in a melt and at

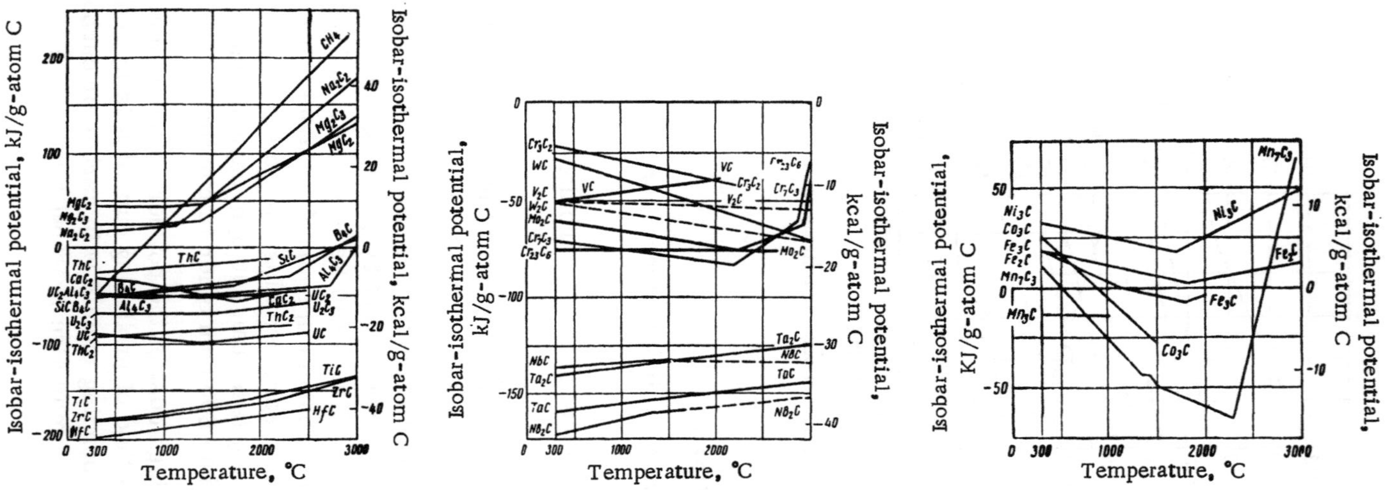

a

b

c

Fig. 23. Temperature dependence of the isobar-isothermal potential of carbide-formation reactions. a) Carbides of metals of groups I, II, III, and IVa, and uranium; b) carbides of metals of subgroups Va and VIa; c) carbides of metals of group VIII and manganese.

temperatures below the melting point of the respective carbides. The most extensively used method is the reduction of metal oxides with carbon at temperatures below the melting points of the carbides.

The reaction of metal oxides with carbon is due to the fact that in a number of metals (particularly the transition metals of groups IV-VI) their affinity for oxygen is less than that of carbon for oxygen (Fig. 22), and hence they can be reduced by carbon at high temperatures and form carbides when there is a fairly great affinity of carbon for the metals (Fig. 23).

The results of the first investigations into the conditions for obtaining carbides are given in [24-26]. However, in spite of the fact that the reduction of oxides with carbon is one of the commonest types of metallurgical processes, for a long time erroneous views regarding its mechanism were widespread.

The reduction of metal oxides with carbon was regarded as a two-stage process [23, 27] consisting of the dissociation of the oxide into metal and oxygen and of the subsequent reaction of the oxygen with carbon:

$$2Me\mathrm{O} = 2Me + \mathrm{O_2};$$
$$\mathrm{C} + \mathrm{O_2} = \mathrm{CO_2}.$$

This view of the reduction process was not supported by the experimental data, i.e., the observed rate of the reduction reaction was considerably higher than what it should have been for the dissociation of even the most unstable oxides.

Another erroneous view of the reduction process was developed in [28-32]. According to this, the reaction proceeds by direct interaction between the components in the solid phase, i.e., by the diffusion of carbon into the oxide. The authors arrived at this conclusion because the experimental data agreed very well with the diffusion law $v^2 = k\tau$.

Moreover, in studying the reduction of the oxides of copper, silicon, manganese, chromium, iron, nickel, and cobalt, Baukloh and his coworkers observed that the gaseous products of the reaction were CO and CO_2, the composition of these products being constant for each particular oxide. From this the following conclusion was reached regarding the dependence of the composition of the gaseous phase on the dissociation pressure of the oxide: the greater the dissociation pressure, the higher the CO_2 content in the gas mixture. The views of Tammann and Baukloh were subjected to criticism [23, 33-38], and their inaccuracy has been demonstrated.

The actual rate of the processes concerned is 100 times and sometimes even 1000 times greater than the rate calculated on the basis of diffusion in the solid phase.

The high rates of the actual reduction processes can be explained only by the decisive part played by the gaseous products which may be formed in the process.

Another of Baukloh's conclusions has also been shown to be invalid, namely that regarding the dependence of the composition of the gaseous products of the reaction on the dissociation pressure of the initial oxides, although the authors' own experiments confirm this.

Thus it is known that $p_{O_2(Fe_3O_4)} > p_{O_2(Cr_2O_3)}$, while according to the authors' data on the reduction of chromium oxide the CO_2 content in the gaseous phase is 80%, whereas on the reduction of Fe_3O_4 it is 65%.

The true mechanism of the reduction of metal oxides by carbon has been formulated by Baikov [33]; according to him, reduction takes place through the agency of carbon monoxide; in this case the principle of the sequence of transformations is observed, by which reduction proceeds from the highest oxides progressively through the formation of lower oxides.

The carbon dioxide formed during the process reacts with carbon and is converted into carbon monoxide in accordance with the equations

$$\begin{array}{c} Me\mathrm{O} + \mathrm{CO} \rightleftarrows Me + \mathrm{CO_2} \\ \mathrm{C} + \mathrm{CO_2} \rightleftarrows 2\mathrm{CO} \\ \hline Me\mathrm{O} + \mathrm{C} \rightleftarrows Me + \mathrm{CO} \end{array}$$

Investigations into the reduction of the oxides of niobium [39], vanadium [40], chromium [41], zinc, and manganese [38] have shown that the gaseous product of the reaction is always CO, and also that the rate of the process depends on the temperature, activity of the reducing agent, particle size of the reducing agent, and pressure in the system and does not depend on the pressure of compaction in briquetting the initial components, i.e., on the extent of the contact between the reacting substances. All this confirms that the reaction proceeds in accordance with the above equations.

Calculation of the activation energy for the reduction of various oxides has given values which are very close to one another, all the values being similar to the activation energy of the reaction $CO_2 + C = 2CO$. From this it follows that the rate at which the reduction of metal oxides by carbon takes place is limited by that of the slowest stage, namely the regeneration of carbon monoxide. However, this cannot be assumed for the reduction of all oxides. In the reduction of the oxides of iron, manganese, chromium, and a number of other very stable oxides, the activation energy is higher than that for the regeneration of CO. In these cases the limiting stage in the initial reduction period is the crystallochemical reorganization of the initial oxides. Then the part played by the gasification of carbon increases, and the reaction is sensitive to all the parameters affecting the rate of CO regeneration. Toward the end of the process, when the surface of the reaction zone contracts, the part played by this factor again diminishes. Thus, the reduction of these oxides is autocatalytic in character — the reduction rate increases rapidly in the initial period, reaches a maximum, and then slowly falls.

According to data in [26], with a hydrogen atmosphere in the furnace the hydrogen reacts with the heated furnace walls in accordance with the equations

$$nC + m/2\,H_2 \rightarrow C_nH_m,$$
$$MeO + C_nH_m \rightarrow MeC + H_2 + H_2O.$$

In addition to the gaseous products indicated, the presence in the gaseous phase of chlorine, hydrogen halides, and chlorine derivatives of hydrocarbons affects the reduction process. These additions accelerate the process of carbide formation [298].

If the reduction of oxides with carbon to obtain carbides is carried out in vacuum, this excludes the possibility of oxidation and nitridation, facilitates the evaporation of volatile impurities, and increases the rate and completeness of the reaction; under these circumstances the equilibrium of the reaction $MeO + C \rightleftharpoons MeC + CO$ is shifted to the right [43].

Kroll [44] determined the temperatures at which the reduction of a number of oxides begins in vacuum; these are considerably lower than the temperatures at which reduction begins at atmospheric pressure. It was also shown that the reaction proceeds through a number of reduction stages to the lowest oxides. This has been confirmed by manometric studies of the reduction of transition-metal oxides by carbon [44-48].

The method of producing carbides by the reaction between metals or oxides and carbon-containing gases is carried out at a sufficiently high partial pressure of the hydrocarbons to provide the quantity of carbon required for carbide formation, but to prevent the precipitation of free carbon.

The reaction of metals with methane and carbon monoxide and also with other carbon-containing gases has been very thoroughly investigated in [49-51]. From the results of these investigations it has been possible to establish the optimum temperatures of the reactions given below:

$Mo + H_2 + CH_4 \rightarrow MoC + H + (CH)$	700
$Mo + H_2 + CH_4 \rightarrow Mo_2C + H_2 + (CH)$	800
$W + H_2 + C_xH_y^* \rightarrow W_2C + H_2 + (CH)$	2100—2400
$W + H_2 + C_xH_y \rightarrow W_2C + H_2 + (CH)$	2440—2550
$W + N_2 + H_2 + C_xH_y \rightarrow WC + H_2 + N_2 + (CH)$	1000—2200
$Nb + H_2 + C_xH_y \rightarrow NbC + H_2 + (CH)$	1300
$Ta + H_2 + C_xH_y \rightarrow TaC + H_2 + (CH)$	1300—2900
$Cr + H_2 + CH_4 \rightarrow Cr_3C_2 + H_2 + (CH)$	600—800

This method takes place by the direct interaction of metal powders with carbon-containing gases and also by deposition on metal wire.

The method is not very much used for the commercial production of carbides. It is mainly used to obtain the carbides of iron, nickel, and cobalt. The mechanism of reaction of the metals, and particularly that of their oxides, with carbon-containing gases has not been thoroughly investigated.

The deposition of carbides from the gaseous phase is one of the most promising methods of producing high-purity compounds. It is based on the reaction between the components of a gaseous mixture consisting of a halide compound of the metal and carbon-

* CH_4, C_6H_6, $C_6H_5CH_3$, C_2H_2, $CO + H_2$, etc.

containing gases [26, 52-54]. Deposition takes place on the surface of an incandescent carbon or refractory-metal filament. Hydrogen promotes the reaction and in certain cases considerably reduces the temperature at which the halide decomposes. This method was first used by van Arkel [52, 54] to obtain zirconium carbide by the reaction

$$ZrCl_4 + CH_4 + H_2 \rightleftarrows ZrC + 4HCl + H_2.$$

The filament temperature and the ratio of the concentrations of the reacting substances have a great effect on the progress of the reaction between the metal halides and the carbon-containing gases. The filament temperature during the reaction should exceed the melting point of the metallic component, and the partial pressures of the reagents should be such as to exclude the possibility of free carbon being deposited. The form of the deposit is determined by the filament temperature: the higher the temperature, the coarser are the deposited crystals. At constant temperature, deposition proceeds by the formation of a layer with uniform cross section. By choosing certain deposition conditions, the carbide can be obtained in the form of a single crystal. It is also possible to deposit carbides from the vapors of the corresponding carbonyl compounds. A study of this process has shown [50] that molybdenum and tungsten carbides can be deposited by the reaction of the respective carbonyl vapors with hydrogen at a temperature of 300-800°C and a pressure of 13.33-39.99 N/m^2 (0.1-0.3 mm Hg) for Mo_2C and 1333.3 N/m^2 (10 mm Hg) for W_2C.

The chemical precipitation of carbides is based on a difference in the solubility of carbides and metallic alloys. This method has been thoroughly investigated in [55-59, 61, 62]. The carbide phases are precipitated by treatment with various chemical reagents in which the metals and their alloys are soluble, while the carbides remain in the form of an insoluble residue, and by the method of anodic solution.

The method of producing carbides by the electrolysis of molten salts was first employed by Andrieux and Weiss [63] to obtain molybdenum and tungsten carbides. It is based on the fact that under certain conditions in the electrolysis of molten carbonates an amorphous carbon deposit is precipitated, which reacts with the metal in solution in the electrolytic bath to form a carbide. The most effective baths are those containing the carbonates, borates, and fluorides of the alkali metals and the oxide of the refractory metal. The introduction of boric acid in the form of sodium metaborate into the electrolyte raises the solubility of the metal oxides in the electrolytic bath.

Chapter III

Carbides of Metals of Group I

Carbides of the Alkali Metals

The System Lithium — Carbon. A study of the lithium — lithium carbide system [4] has revealed the existence of a single carbide phase Li_2C_2, which forms a eutectic with lithium at 165°C and 0.3 at.% C. Lithium carbide is a white or grayish substance which crystallizes in the rhombic lattice [241, 262].

The results of thermal analysis and x-ray examination have shown that there are four polymorphic modifications of lithium carbide, the transformation temperatures occurring at 410, 440, and 560°C [66].

The first version of part of the phase diagram of the Li — Li_2C_2 system is shown in Fig. 24 [4].

Lithium carbide Li_2C_2 was first obtained by Moissan, who reduced lithium carbonate with carbon: $Li_2CO_3 + 4C = Li_2C_2 + 3CO$ [64]. Subsequent investigations showed that the carbide can also be produced by heating a mixture of the metal and carbon in a vacuum at 600-800°C [212, 241]. At higher temperatures, the carbide easily dissociates. At 700-800°C, lithium reacts with carbon monoxide and hydrocarbons to form the carbide.

Heating lithium in an ethylene atmosphere leads to the formation of a molten grayish-white product, which, according to analysis, corresponds to the composition $C_2H_4Li_6$. Since the formation of such a compound is unlikely, it has been shown that the

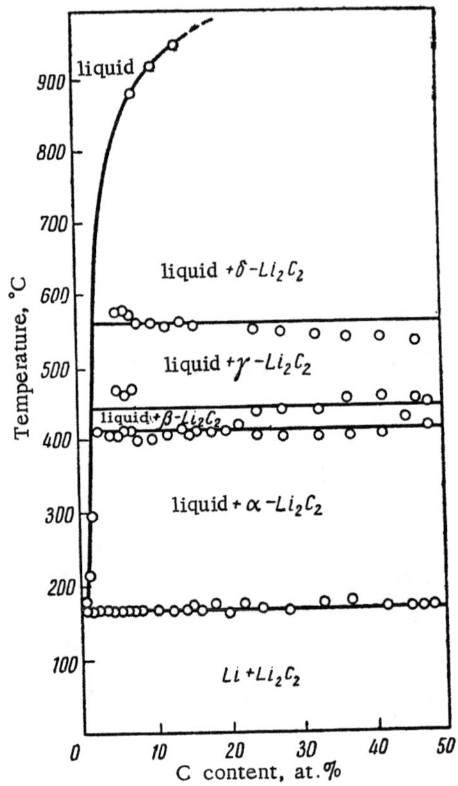

Fig. 24. Phase diagram of $Li - Li_2C_2$ system [4].

following reaction takes place:

$$6Li + C_2H_4 = Li_2C_2 + 4LiH.$$

On decomposing the reaction product with water, acetylene and hydrogen were given off in quantities corresponding to the reaction:

$$Li_2C_2 + 4LiH + 6H_2O = C_2H_2 + 4H_2 + 6LiOH.$$

When lithium was slowly heated in an iron boat in a stream of acetylene, carbon was precipitated with simultaneous formation of lithium carbide and hydride.

Lithium carbide has also been obtained by the action of acetylene on a solution of lithium in ammonia [64]. In this case, a double compound $Li_2C_2 \cdot C_2H_2 \cdot 2NH_3$ was formed, which decomposed in a stream of hydrogen at room temperature with the precipitation of finely dispersed white lithium carbide powder.

The System Sodium – Carbon. No reaction between sodium and carbon is observed at temperatures up to 800°C [5, 64]. It was shown in [67] that the reaction between sodium vapor and carbon depends on the kind of carbon-containing material used. Thus, while sodium vapor scarcely reacts at all with graphite, its reaction with carbon black proceeds fairly rapidly. Three phases – Na_2C_2, NaC_8, and NaC_{16} – have been found in the sodium – carbon system [71]. Of these, Na_2C_2 has been most thoroughly investigated; it is a white powder with a density of 1.57 g/cm^3 at 15°C. It dissociates at 800°C [64].

The principal method of producing sodium carbide is by the reaction between metallic sodium and acetylene. At temperatures between the melting point of sodium and 190°C, the reaction proceeds as follows:

$$2HC{\equiv}CH+2Na=2NaC{\equiv}CH+H_2+28.9\text{ kcal }(121.1\text{ kJ}).$$

At 200°C, the acid salt decomposed [68]:

$$2NaC_2H = Na_2C_2 + C_2H_2.$$

At 400-600°C sodium carbide is formed directly by the reaction

$$C_2H_2 + 2Na = Na_2C_2 + H_2 + 206.6\text{ kJ }(49.3\text{ kcal}).$$

Sodium carbide has also been obtained [69] by passing dry acetylene through a solution of sodium in ammonia. In this case, sodium acetylide is produced,

$$3C_2H_2 + 2NH_3Na = 2NaHC_2 + 2NH_3 + C_2H_4,$$

which decomposes on heating with the precipitation of Na_2C_2.

Sodium carbide is also formed by the interaction of sodium with organic compounds [70].

The System Potassium – Carbon. While investigating the reaction products of potassium with carbon on heating in a hydrogen atmosphere, Dewey [64] observed the formation of a dark, electrically conducting powder which readily burned in air. This compound corresponded to the formula K_2C_2.

In investigations into the reaction between potassium vapor and carbon at high temperatures [67, 73, 74], it was found that graphite dissolves in liquid potassium even at 50°C, its solubility increasing with rise in temperature. At 300-400°C, potassium vapor reacts with carbon to form products containing 4, 8, and 16 moles of carbon per mole of potassium. The compound KC_8 forms coarse, copper-red crystals with the graphite lattice, in which the potassium atoms are situated in the spaces between the layers. The steel-blue compound KC_{16} has a metallic luster. The potassium atoms are again situated in the spaces between the layers of carbon atoms. There are data in the literature on the carbides KC_{24}, KC_{36}, KC_{48}, and KC_{60} [10].

The most widely used method of producing potassium carbide is by the interaction of potassium with acetylene [75]. Depending on the temperature, either K_2C_2 or KHC_2 may be obtained; the latter is formed at 25-30°C or by passing acetylene through a solution of potassium in liquid ammonia. It is a colorless crystalline powder (density 1.37 g/cm^3), which gives off acetylene at high temperatures and is transformed into K_2C_2. The structure of KHC_2 is similar in type to that of CaC_2 [72]. The carbide K_2C_2 is also obtained by the reaction between acetylene and potassium dispersed in acid at 120-130°C [5]. After washing the deposit precipitated by the reaction with dry benzene or ether and then drying it, a yellow powder is obtained which contains up to 70% of K_2C_2.

Potassium carbide may also be produced at 300°C by the reactions

$$6KOCH_3 \rightarrow K_2C_2 + 2K_2CO_3 + 2C + 9H_2$$

and

$$2KCN + 3Mg = K_2C_2 + Mg_3N_2.$$

The System Rubidium—Carbon. Rubidium carbide was first obtained by the action of acetylene on an ammoniacal solution of rubidium [76]. The product was an acetylene compound $Rb_2C_2 \cdot C_2H_2$, which on rapid heating in vacuum up to 300°C decomposed with the precipitation of a pure dark-red powder Rb_2C_2. When $Rb_2C_2 \cdot C_2H_2$ was slowly heated in vacuum, carbon, metal, and poly-

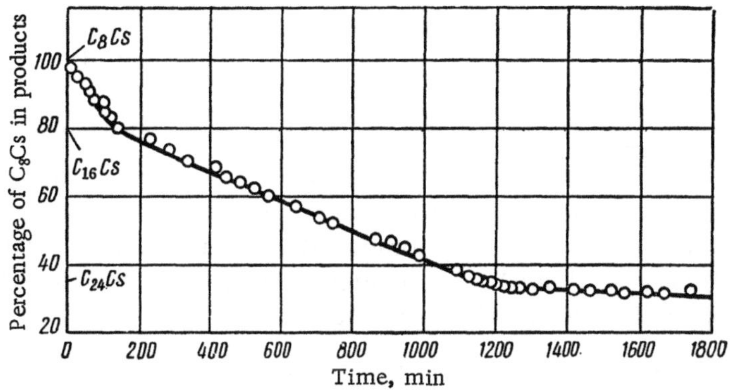

Fig. 25. Dependence of rate of decomposition of CsC_8 on time.

merization products of acetylene were formed. As in the case of potassium, compounds richer in carbon — RbC_8, RbC_{24}, RbC_{36}, RbC_{48}, RbC_{60} — are formed by the action of rubidium vapor on graphite [10].

<u>The System Cesium—Carbon.</u> Cesium carbide has been obtained by passing acetylene through an ammoniacal solution of cesium [76]. The solution rapidly lost color, cesium monoacetylide was formed as a white crystalline powder, and ethylene was given off.

On rapid heating in vacuum at 300°C, the acetylide $Cs_2C_2 \cdot C_2H_2$ decomposes with the formation of the carbide Cs_2C_2 and evolution of acetylene. On slow heating it decomposes into the metal, amorphous carbon, and polymerization products of acetylene. On heating in vacuum, Cs_2C_2 decomposes into the metal and carbon at 700–800°C [67].

The carbide CsC_8 has been obtained by the reaction of pyrolytic graphite with cesium vapor at a temperature of 300°C and a pressure of 53.3 N/m^2 (0.4 mm Hg) [10].

It has been shown that the carbide CsC_8 decomposes at 400°C, the time dependence of the decomposition rate being as expressed by the curve in Fig. 25, in which there are two breaks: one at the transition to the CsC_{16} phase and the other at the transition from CsC_{16} to CsC_{24}.

Data are given on the existence of the carbides CsC_{36}, CsC_{48}, and CsC_{60}, whose properties have also been investigated.

Carbides of Metals of the Copper Subgroup

The System Copper—Carbon. An investigation of the solubility of carbon in liquid copper has shown [77] that it is 0.0001% at 1100°C, increasing somewhat with rise in temperature to reach 0.003% at 1700°C. According to [78], the solubility of carbon in copper at 2200–2300°C is 0.024–0.033%.

The existence of the copper carbides Cu_2C and Cu_2C_2 has been established.

X-ray examination of Cu_2C_2 has shown it to have a structure similar to that of calcium carbide with a face-centered tetragonal lattice. The pycnometric density of Cu_2C_2 is 4.62 g/cm^3.

Cu_2C_2 decomposes explosively on contact with the sharp edges of a solid substance [80].

It has not proved possible to obtain the carbide by the direct reaction of copper with carbon.

On passing acetylene through an ammoniacal solution of cupric chloride, a black precipitate is formed whose composition is close to that of Cu_2C [66]. It decomposes at 70–80°C.

The carbide Cu_2C_2 is the product of the replacement of hydrogen in acetylene by copper; it is formed as a dark-red crystalline substance by the reaction between an ammoniacal solution of CuI and KHC_2 in the ratio 1 : 1 or 1 : 2 [79].

The System Silver—Carbon. The solubility of carbon in silver is 0.0012–0.0025% in the range 1600–1940°C [78]. On cooling, all of the carbon is precipitated in the form of graphite from the solution. Acetylene precipitates grayish-yellow silver carbide from an ammoniacal solution of silver nitrate. The dried Ag_2C_2 decomposes explosively on heating [66].

The System Gold—Carbon. According to [81], gold dissolves only a negligible amount of carbon. The carbide Au_2C_2 is formed as a yellow precipitate by the reaction of acetylene with gold sodium dithiosulfate [82]; after washing with water and alcohol and filtering, it is dried over sulfuric acid. In the dry state it is an explosive substance.

Chapter IV

Carbides of Metals of Group II

Carbides of Beryllium, Magnesium, and the Alkaline-Earth Metals

The System Beryllium — Carbon. The first investigations of the system Be — C [84-86] revealed the existence of two carbides: Be_2C and BeC_2. However, very few data are available in the literature on the higher carbide BeC_2. The solubility of carbon in beryllium is negligible.

The carbide Be_2C (39.98% C) is a reddish crystalline substance which is very hard and brittle [89, 93, 94].

A study of the physical properties of hot-compacted specimens of beryllium carbide containing various amounts of graphite [88] has shown that the addition of graphite, while facilitating the production of dense specimens, has no marked effect on the temperature coefficient of linear expansion, nor does it affect the thermal conductivity or the specific electrical resistivity. The specific resistivity decreases with rise in temperature in all mixtures containing up to 50% graphite. The mixture of 50% Be_2C and 50% C has properties similar to those of graphite.

By a manometric study [91] of the equilibrium of the reaction between beryllium oxide and carbon ($BeO_{solid} + {}^3/_2C_{solid} = {}^1/_2Be_2C_{solid} + CO_{gas}$), it has been found that the equilibrium state corresponding to the reaction is attained in the temperature range 1400-1650°C. The reduction products contain beryllium carbide and unreacted beryllium oxide and carbon.

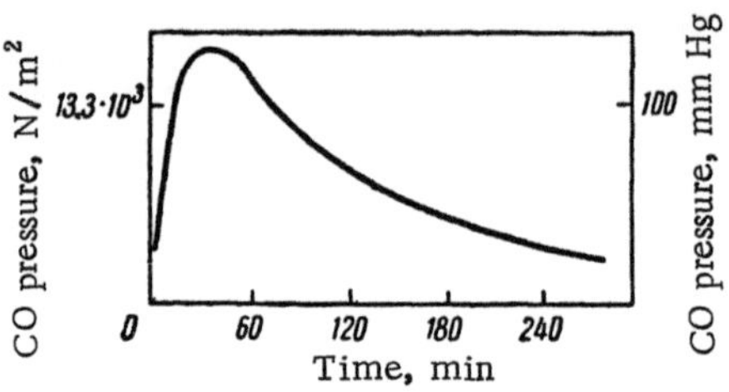

Fig. 26. Dependence of CO pressure on time in the reaction $BeO + C \rightarrow Be_2C + CO$.

Figure 26 shows the time dependence of the carbon monoxide pressure at 1890°C. The reaction products corresponding to the rising part of the curve contain Be_2C, BeO, and C; those on the falling part of the curve consist of almost pure carbon. Evidently, beryllium carbide dissociates above 1680°C: $Be_2C_{solid} = 2Be_{gas} + C_{solid}$; while in the presence of the gaseous phase CO: $Be_{gas} + CO = BeO_{solid} + C_{solid}$, which leads to the deposition of berylium oxide as well as carbon on the cold parts of the heater.

The equilibrium dissociation pressure of Be_2C in accordance with the reaction $\frac{1}{2}Be_2C_{solid} \rightleftharpoons Be_{gas} + \frac{1}{2}C_{solid}$, which has been investigated in the range 1430-1669°K by Knudsen's method [92], corresponds to the equation

$$\log p = (7.026 \pm 0.347) - \frac{19720 \pm 537}{T}.$$

The carbide Be_2C has been obtained [88] by melting beryllium in graphite crucibles and subsequently treating the reaction products with hydrochloric acid. The efficiency of this method is low, and it is used only for producing small amounts of the carbide.

A second method of obtaining beryllium carbide is by reducing the oxide with carbon. In [87], this process was carried out in a vacuum electric furnace at 1900°C. The yield was 60%.

A product containing 85-92% Be_2C has been obtained [89] by compacting a mixture of BeO and wood charcoal into briquets which were then heated in vacuum at 1900°C.

On a commercial scale, Be_2C has been produced [88] by means of a charge of BeO and graphite containing 1.5% excess graphite and 2% paraffin. The charge was compacted into briquets which were heated in a graphite crucible in an induction furnace at 1950–2100°C for 8 h in a purified hydrogen atmosphere.

The conditions for producing dense specimens of beryllium carbide were investigated in the same paper [88]. The compacted powders containing metallic additions (10% Fe, Re, Si) were sintered in a protective atmosphere at 2250–2300°C for 30 min. The specimens obtained were dense and not brittle. Hard, dense specimens were also obtained by hot compaction at 1900–2250°C in a purified hydrogen atmosphere.

Experiments carried out in hydrogen, helium, and argon atmospheres showed hydrogen to be the most effective.

The System Magnesium—Carbon. Two carbides are known in the Mg—C system: Mg_2C_3 (42.56% C) and MgC_2 (49.69% C) [85].

MgC_2 can be produced by heating magnesium powder in a stream of acetylene or other hydrocarbons at 500°C [99]. Mg_2C_3 is formed by the action of pentane on metallic magnesium at 650°C. MgC_2 is not formed under these conditions.

An investigation into the possibility of obtaining MgC_2 by the reaction of acetylene with magnesium ethyl bromide [100] has shown that the magnesium acetylene dibromide formed in this way decomposes on heating into $MgBr_2$ and MgC_2, which also breaks down on heating:

$$2MgC_2 \rightleftarrows Mg_2C_3 + C.$$

MgC_2 decomposes at 600°C. Mg_2C_3 decomposes at 740–760°C:

$$Mg_2C_3 \rightleftarrows 2Mg_{gas} + 3C.$$

The conditions for producing MgC_2 and Mg_2C_3 have been very thoroughly investigated in [101].

The reaction between $MgCl_2$ and CaC_2 and between a mixture of $MgCl_2$ + NaCl and CaC_2 has been studied at 600–700°C in an argon atmosphere. It was found that the reaction between $MgCl_2$ and CaC_2 at temperatures below 622°C (the $MgCl_2$—$CaCl_2$ eutectic temperature) leads to practically no formation of Mg_2C_3. At higher temperatures the reaction passes through a stage of MgC_2 formation. The reaction rate reaches a maximum at 600°C. The thermal stability boundary of Mg_2C_3 occurs at 740°C, in good agreement with the data in [99, 100]. The function of NaCl in the reaction is to lower the melting point of $MgCl_2$.

According to [102], magnesium carbide can be obtained by the electrolysis of a molten salt mixture consisting of (mol.%): KCl, 20; NaCl, 32; and $MgCl_2$, 48, using carbon cathodes and molybdenum anodes.

The System Calcium–Carbon. Only one carbide phase (CaC_2) is known to exist in the Ca–C system. CaC_2 (37.48% C) forms colorless crystals with a density in the pure form of 2.22 g/cm^3.

It has been established in numerous researches [93, 117-122, 125] that CaC_2 exists in four modifications. The most extensively studied of these are CaC_2-I, with an fc-tetragonal lattice containing four CaC_2 molecules per unit cell; it is stable up to 447°C, and has a coordination number (z) of Ca in relation to C of 10; and CaC_2-II, which is stable above 447°C, and crystallizes in triclinic syngony with parameters $a = 0.842$, $b = 1.148$, and $c = 0.394$ nm; $\alpha = 93.4°$, $\beta = 92.5°$, and $\gamma = 89.9°$; z = 8.

CaC_2-II has been obtained by heating commercial CaC_2 with $CaCN_2$ and graphite at 1700°C [112].

According to [122], CaC_2-III has a monoclinic structure, while CaC_2-IV has an fcc structure.

Ordinary commercial calcium carbide consists of the tetragonal modification CaC_2-I.

The electrical conductivity of calcium carbide [124] depends on its chemical composition and structure and also on its thermal treatment [123]. Anisotropy occurs as a result of the different cooling conditions in the various parts of the ingot. The outer layers, which cool most rapidly, have a higher specific electrical resistivity than the inner layers, which cool more slowly.

Annealing leads to a reduction in the resistivity owing to the transformation of CaC_2-I into another modification (CaC_2-II or CaC_2-III) both of which have polymerized anions $(C_2^{2-})_m$ or $(C_2O^{4-})_n$ which confer semiconducting properties on the compound.

The heat of formation of CaC_2 has been determined by indirect methods in a number of researches. Thus, from a determination of the heat of solution of CaC_2 and Ca in dilute HCl in accordance with the reactions

$$\begin{array}{c} CaC_2 + 2HCl = CaCl_2 + C_2H_2 + X \text{ J} \\ Ca + 2HCl = CaCl_2 + H_2 + 540.1 \text{ J} \\ \hline CaC_2 + H_2 = Ca + C_2H_2 + X - 540.1 \text{ J} \\ 2C + H_2 = C_2H_2 - 225.8 \text{ J} \\ \hline Ca + 2C = CaC_2 + (314.3 - X) \text{ J} \end{array}$$

the heat of formation of CaC_2 has been found to be 59.1 kJ/mole (14.1-14.2 kcal/mole) [126, 273]. The thermodynamic characteristics of CaC_2 are given in [60, 103].

CaC_2 dissociates on heating in vacuum, and this fact is made use of to obtain pure calcium [137, 138]. The dissociation pressure of CaC_2 is 133.3 N/m^2 (1 mm Hg) at 1850°C and 101.3 kN/m^2 (760 mm Hg) at 2500°C.

In studying the dissociation of CaC_2 in vacuum, Gel'd and Esin [139] found that this reaction is markedly activated by salt additions. The part played by the additions is that the anions of the adsorbed salts interact with the calcium and weaken its bond with carbon. Since fluorine ions react more strongly with calcium than do chlorine ions, the accelerating action of CaF_2 on the dissociation of CaC_2 is greater than that of $CaCl_2$.

According to [104], CaC_2 can be obtained by heating calcium, purified by distillation in argon at a pressure of 399.9 N/m^2 (3 mm Hg), with carbon black heated in vacuum at 1500°C, in an argon atmosphere at 1250°C. The products derived from this process are remelted under argon, the composition of the final products being: 93-94% CaC_2, up to 2% CaO, and up to 2% C_{free}.

The method most widely used at present for producing CaC_2 is by the reduction of CaO with carbon ($CaO + 3C = CaC_2 + CO$) at 1840-2070°C; this method has been extensively described in the literature [64, 104-115].

Anthracite, coke, and wood charcoal are used as the carbonaceous materials. The principal requirement in regard to these materials is that they shall contain as little gas and volatile substances as possible. The most suitable material in this respect is anthracite, which has the highest carbon content of the materials mentioned and the lowest content of ash and volatile impurities.

The purest product, containing 99% CaC_2, is made by heating a mixture of calcium cyanamide with carbon in high vacuum at 1100–1150°C, and then raising the temperature to 1350°C. Because CaC_2 is easily decomposed in air, it decomposes appreciably (3–10%) during cooling and crushing in manufacture. The behavior of CaC_2 in various gaseous media was investigated in [845] with the object of finding an atmosphere which prevents decomposition. It was shown that at temperatures above 800°C such atmospheres as nitrogen, oxygen, air freed from CO_2 and moisture, CO, and CO_2 are aggressive and cause the CaC_2 to decompose. The extent of decomposition can be reduced only by more drastic cooling in the temperature range 1000–1200°C.

The System Strontium—Carbon. Strontium carbide SrC_2 has been obtained [64] by heating a mixture of strontium oxide or carbonate with carbon.

An x-ray examination has shown [93] that the product usually obtained contains a large amount of impurities in the form of SrO and graphite.

Three modifications of SrC_2 are known to exist: a cubic one, stable above 370°C, with a lattice parameter a = 0.625 nm; an fc-tetragonal one with parameters a = 0.582 and c = 0.669 nm, c/a = 1.150, which is stable in the temperature range from −90 to +370°C; and a modification stable below −90°C.

The System Barium—Carbon. BaC_2 is the only phase to have been found in the Ba—C system; at room temperature it crystallizes in a CaC_2-type tetragonal lattice with parameters a = 0.622, c = 0.706 nm, c/a = 1.134 [93, 119]; above 150°C it crystallizes in an fcc lattice [85].

Like beryllium and calcium carbides, BaC_2 dissociates into the metal and carbon on heating in vacuum at 1355–1671°C:

$$BaC_{2\,solid} = Ba_{gas} + 2C_{solid}$$

Data on the dissociation rate of BaC_2 and the dissociation pressure of barium vapor over the carbide have shown that the rate increases with rise in temperature from $4.5 \cdot 10^{-4}$ at 1355°C to $1.5 \cdot 10^{-2}$ g/cm$^2 \cdot$ sec at 1671°C, and that the temperature dependence of the dissociation pressure is expressed by the equation [151]

$$\log p = 3.267 - \frac{10660}{T}.$$

BaC_2 was first obtained by heating a 20% barium amalgam with powdered carbon in a hydrogen atmosphere at 700-800°C [64]. Subsequently, a mixture of barium carbide and magnesium oxide was obtained by heating a mixture of barium carbonate and magnesium in the presence of carbon black at 700-800°C:

$$BaCO_3 + 3Mg + C = BaC_2 + 3MgO.$$

The BaC_2 content in the mixture was not more than 38%. However, it proved impossible to separate the carbide from the MgO.

A study of the reduction of BaO with carbon black has shown [146, 147] that the reaction proceeds in vacuum at 1400°C with the evolution of CO and the formation of a carbide — oxide mixture.

Investigations of the reactions

$$BaO + 3C = BaC_2 + CO$$

and

$$BaO_2 + 4C = BaC_2 + 2CO$$

have shown [148] that it is practically impossible to obtain BaC_2 by reducing BaO with carbon in vacuum at 1000-1500°C, owing to the high volatility of the barium oxides. BaC_2 is formed by the reduction of barium carbonate with carbon; in this reaction it must be borne in mind that the dissociation pressure of $BaCO_3$ is so high at 1300-1400°C that it decomposes almost completely into BaO and CO_2, and that the dissociation rate of $BaCO_3$ is greater than the rate of regeneration of CO by the reaction

$$CO_2 + C = 2CO.$$

In summary, the reaction proceeds in accordance with the equation

$$BaCO_3 + 3C = BaC_2 + CO_2 + CO.$$

This conclusion is confirmed by experimental data which show that at 1600°C and with a charge containing 5% excess carbon black over the theoretical requirement, BaC_2 is formed which has a composition close to the stoichiometric ($C_{comb} = 14.0\%$).

Carbides of the Zinc Subgroup

The System Zinc—Carbon. Boiling zinc dissolves only a negligible amount of carbon [78]. No compounds of zinc with carbon are known. A number of papers deal with the investigation of the thermodynamics [154, 155] and kinetics [156-158] of the reduction of ZnO with carbon and CO. The papers show that the reduction leads to the formation of metallic zinc.

Although zinc does not form individual carbides, the following fairly stable ternary carbide alloys are known: Ti_2ZnC_x, Zr_2ZnC_x, Hf_2ZnC_x, Nb_2ZnC_x [159, 305], and $Mn_{60}Zn_{20}C_{20}$ [160].

Like zinc, cadmium and mercury form no compounds with carbon.

Chapter V

Carbides of the Transition Metals

Carbides of Scandium, Yttrium, and the Lanthanides

The System Scandium–Carbon. According to the first investigations, a single carbide corresponding in composition to the formula Sc_4C_3 exists in the system Sc — C [25]. However, subsequent investigations have not confirmed the existence of this carbide. A carbide has been obtained corresponding to the formula ScC, which crystallizes in the NaCl-type cubic lattice [179], and the existence of the carbide Sc_2C with a hexagonal structure has been suggested. Vickery and his coworkers [180] have confirmed the existence of the carbide ScC and found that it has a hexagonal structure; however, this has not been confirmed in subsequent investigations and is evidently incorrect.

It has been shown, according to x-ray data, in [181, 182] that on reducing Sc_2O_3 with carbon at temperatures above 2000° C in a protective gas atmosphere, the reaction products contain, in addition to Sc_2O_3 and graphite, a cubic phase with a NaCl-type structure and a parameter a = 0.447 nm. It has been concluded that either a scandium carbide with a deficiency of carbon atoms, ScC_{1-x}, or an oxycarbide Sc(C, O) is formed. The formation of oxycarbides has been confirmed in [183, 184]. Consequently, there is in the system Sc — C a carbide closely similar in composition to ScC, which crystallizes in a cubic structure and tends to form oxycarbides.

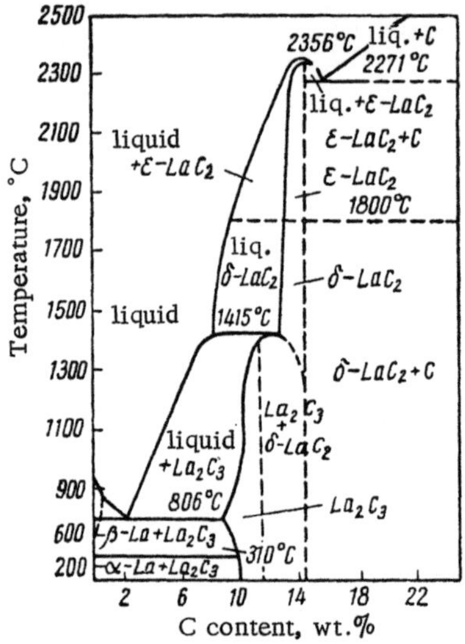

Fig. 27. Phase diagram of the system La — C.

<u>The System Yttrium—Carbon.</u> The existence of four phases has been established in the system Y — C: Y_3C, YC, Y_2C_3, and YC_2. The carbide Y_3C_{1-x}, which is lowest in carbon content, lies in the composition range from $Y_3C_{0.75}$ to $Y_3C_{1.2}$.

A study of certain properties of yttrium carbides having the compositions $Y_3C_{0.9}$ and $Y_3C_{1.11}$ has shown [864] that the lattice parameters are 0.5068 ± 10^{-4} nm for $Y_3C_{0.9}$ and 0.506 nm for $Y_3C_{1.11}$; the specific electrical resistivities are 130 and 160 $\mu\Omega \cdot$ cm, respectively, and the molar magnetic susceptibilities $256 \cdot 10^{-6}$ and $126 \cdot 10^{-6}$ CGS units.

The monocarbide YC is obtained by the reduction of yttrium oxide with carbon in vacuum at 1900°C. A study of the conditions for obtaining YC has shown [185, 186] that yttrium oxycarbide is formed at 1700°C; with rise in temperature, oxygen is removed from the oxycarbide as CO with the formation of YC at 1900°C.

Above 1900°C the monocarbide melts, decomposing into yttrium, which is partly evaporated in vacuum, and a liquid phase rich in carbon.

The sesquicarbide Y_2C_3 is obtained by the reaction of yttrium oxide with carbon in vacuum at 1800°C. As in the case of the monocarbide, the process takes place via the formation of the oxycarbides.

The dicarbide YC_2 was first obtained by heating a fine mixture of yttrium ore with coal in an arc furnace [187, 188]. An investigation of the conditions for producing YC_2 by the reduction of yttrium oxide with carbon in vacuum has shown [185, 186] that the product obtained at 1900°C has a composition close to the calculated one. In contrast to YC and Y_2C_3, the dicarbide is stable in vacuum at 1900°C.

The System Lanthanum–Carbon. According to the phase diagram that has been proposed in [7], there are two carbides in the La–C system (Fig. 27), namely La_2C_3 and LaC_2, of which the former melts incongruently and the latter congruently. The solubility of carbon in solid lanthanum at 695°C is 1.14%. La_2C_3 is homogeneous over the composition range 9.98-11.58% C (the calculated carbon content in La_2C_3 is 11.48%).

La_2C_{3-x} has a bcc lattice, the parameter of which is 0.8803 ± 0.00004 nm at the metal-rich end and 0.8185 ± 0.00004 nm at the carbon-rich end [7].

The homogeneity range of LaC_2 is very small. At 1800°C, LaC_2 undergoes a transformation. A mass-spectrometric study of the evaporation of LaC_2 has shown [170] that at 2200°C there are molecules of the carbide LaC_2 and lanthanum atoms in the vapor above the dicarbide; the ratio of the mass intensity peaks of La ions to LaC_2 is 1:16.

A thorough investigation of the conditions under which the sesquicarbide La_2C_{3-x} is obtained has shown [196] that it is not formed by the reduction of lanthanum oxide with carbon in vacuum or in an inert (argon) atmosphere. This is probably due to the formation of the thermodynamically more stable dicarbide. La_2C_{3-x} is obtained by the reaction of LaC_2 with metallic lanthanum at 1400-1500°C in an argon atmosphere. However, this method is of little use commercially, as it first requires the production of the dicarbide, which is a very unstable compound, easily decomposed by atmospheric moisture.

Fairly pure sesquicarbide can be obtained by synthesis from the elements by arc melting.

An investigation of the conditions for obtaining LaC_2 by the reduction of lanthanum oxide with carbon in vacuum has shown [190-192] that the reaction begins at 1500°C. The optimum temperature for obtaining the dicarbide is 1800-1900°C.

The System Cerium–Carbon. Data on the existence of the phases CeC, Ce_2C_3, and CeC_2 in the Ce–C system are given in [119].

An x-ray study of this system [193] has confirmed the existence of only two of these carbides: Ce_2C_3 and CeC_2. In the opinion of those authors, the phase CeC previously described by Stackelberg is a solid solution of carbon in cerium. However, Dancy et al. [820], in an investigation into the equilibrium of the system Ce — H_2 — CH_4, have confirmed the existence of the carbide CeC, which has an fcc lattice. The existence of CeC_3 described by Werf [7] was not confirmed.

The sesquicarbide Ce_2C_3 is obtained by the reaction of the dicarbide with metallic cerium and by synthesis from the elements using arc melting [196]. The dicarbide was first produced by reducing cerium oxide with carbon [187]. Investigations into the production and properties of CeC_2 are described in [64, 180, 192].

A very thorough investigation into the reduction of cerium oxide with carbon in vacuum in order to obtain CeC_2 has shown that reduction begins at 1300°C and that at 1700–1800°C a product is formed corresponding in composition to CeC_2 [192]. This product is a golden-yellow crystalline powder.

The System Praseodymium — Carbon. X-ray investigations [7] have established the existence of two phases, Pr_2C_3 and PrC_2, in the Pr — C system. The existence of the previously reported phases Pr_3C and PrC was not confirmed.

The production of the monocarbide PrC during an investigation of the system Pr — H_2 — CH_4 has been described in [820]; however, these data require verification.

Praseodymium carbides have been obtained by the reduction of the oxide with carbon in [7, 180]. However, attempts to obtain Pr_2C_3 have shown [196] that the single-phase sesquicarbide cannot be produced by this method. The most effective methods are by arc melting a mixture of praseodymium and carbon and by the interaction of the higher carbide PrC_2 with metallic praseodymium in induction furnaces at 1400–1500°C.

The System Neodymium — Carbon. The phases Nd_2C_3 and NdC_2 have been found in the Nd — C system [7]. The existence of the previously described carbides Nd_3C and NdC was not confirmed. The conditions for obtaining neodymium carbides, and their properties, are similar to those for praseodymium carbides.

The carbide phases of samarium, europium, gadolinium, terbium, dysprosium, holmium, erbium, thulium, ytterbium, and lutecium have not been properly investigated. It is known, however, that these metals form carbides of composition Me_3C (excepting europium) and MeC_2. The existence of Me_2C_3 phases has also been established in certain cases [7, 180].

The sesquicarbides of holmium, erbium, thulium, and lutecium are isostructural with Y_2C_3. According to data in [7], Ho_2C_3 exists in two isostructural modifications having Pr_2C_3 and Y_2C_3 lattices. Yb_2C_3 is not isostructural with either Pr_2C_3 or Y_2C_3. The carbide Ho_2C has been obtained [873] by the interaction of holmium hydride with carbon at 800°C in an argon atmosphere, and a neutron-diffraction study of its structure has been made.

Carbides of the Actinides

The System Thorium—Carbon. According to the phase diagram of the Th—C system proposed in [197, 198], there are two compounds in this system: a monocarbide containing 4.92% C, melting at 2625°C, and a dicarbide containing 9.37% C and melting at 2655°C. At high temperatures these two carbides form a continuous series of solid solutions; at room temperature they are almost completely insoluble in one another.

More recent investigations of the Th—C system [199] have confirmed the existence of the fcc monocarbide and the dicarbide existing in monoclinic and cubic (CaF_2-type) modifications. The proposed phase diagram is given in Fig. 28.

An investigation into the diffusion of carbon into thorium [201] by determining the carbon concentration gradient along the radius of a thorium specimen at various diffusion-annealing times has shown that the diffusion coefficient is $7.4 \cdot 10^{-9}$ cm²/sec at 1000°C, $2.8 \cdot 10^{-8}$ cm²/sec at 1100°C, and $5.7 \cdot 10^{-8}$ cm²/sec at 1200°C. The activation energy of diffusion is 159.2 kJ/mole (38 kcal/mole).

In a study of the solubility of carbon in thorium [202] it has been found from x-ray and metallographic analyses that the solubility of C (wt.%) is 0.35 at room temperature, 0.43 at 800°C, 0.57 at 1018°C, and 0.91 at 1250°C.

The monocarbide crystallizes in a NaCl-type fcc lattice. The data given by various authors for the lattice parameter vary

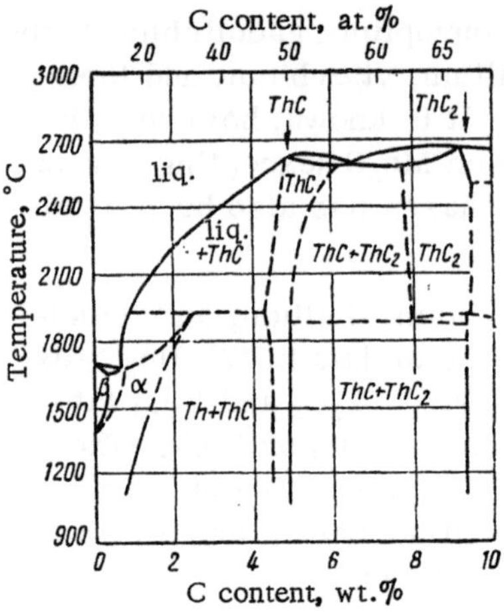

Fig. 28. Phase diagram of the system Th – C [199].

somewhat, the values of *a* being 0.5325 nm [198], 0.529 nm [209], and 0.534 nm [197].

It has been found in [197] that the lattice parameter of ThC varies from 0.529 to 0.534 nm, depending on the composition. Measurement of the lattice parameter of ThC_{1-x} in relation to composition has shown [208] that at $0 < x < 0.4$, i.e., from Th to $ThC_{0.65}$, the lattice parameters do not vary and the products can be regarded as solid solutions of carbon in thorium. In the composition range $ThC_{0.65}$ – $ThC_{0.95}$ the lattice parameter increases monotonically from 0.5296 to 0.5335 nm, and then again becomes constant.

Satow arrived at the same results ($ThC_{0.68}$ – $ThC_{0.95}$) in [952], in which the dependence of the enthalpy and entropy of ThC_x on composition was investigated by the emf method.

The dicarbide crystallizes in an fc-tetragonal lattice with parameters $a = 0.586$ and $c = 0.529$ nm, and with four molecules per unit cell [119]. There are also data [209] to show that it has a bc-tetragonal lattice with parameters $a = 0.2144$ and $c = 0.529$ nm, with two molecules per unit cell, and further data [204] indicating that ThC_{2-x} is pseudotetragonal and crystallizes in an orthorhombic lattice with parameters $a = 0.826$, $b = 1.052$, and $c = 0.422$ nm, with eight molecules per unit cell.

It has been shown in [215] by x-ray analysis and neutron diffraction of single crystals that ThC_{2-x} crystallizes in a monoclinic lattice with parameters $a = 0.653$, $b = 0.424$, $c = 0.656$ nm, $\beta =$ 104°, with four ThC_{2-x} molecules per unit cell. This has been confirmed by data in [134], according to which the lattice parameters of $ThC_{1.99}$ are $a = 0.6691 \pm 0.0001$, $b = 0.4231 \pm 0.0001$, $c = 0.6744 \pm 0.0002$ nm, $\beta = 103°50'$.

The researches described in [134, 215] were carried out by precision methods on pure materials, taking account of their exact composition, and the results may therefore be regarded as very reliable.

At 1900°C, the monoclinic modification of ThC_{2-x} is transformed into a CaF_2-type cubic lattice with parameter a = 0.5808 nm.

Later, in an investigation into the temperature dependence of the electrical resistivity of Th — C alloys containing 4–70 at.% C, it was found [953] that at 1370°C the monoclinic modification of the dicarbide changes into the tetragonal form and this in turn is transformed at 1470°C into the cubic modification.

Some thermodynamic properties of thorium carbides having compositions between $ThC_{1.02}$ and $ThC_{2.43}$ have been determined by the emf method, using cells of the types Th, $ThF_4(CaF_2)ThF_4$, ThC_x [200], and W[Th, $ThF_4(CaF_2)ThF_4$, ThC_x]W [952], and it has been shown that the enthalpy and entropy decrease monotonically on passing from $ThC_{1.72}$ to $ThC_{1.97}$. With x in ThC_x less than 1.72 and greater than 1.97, they are constant. Consequently, the homogeneity range of thorium dicarbide lies between $ThC_{1.72}$ and $ThC_{1.97}$.

From measurements of the vapor pressure of ThC_{2-x} by Knudsen's effusion method and analysis of the condensate by an activation method [211], it has been possible to calculate the heat of formation of the dicarbide (Table 2).

The thermal conductivity of thorium carbides has been measured in [214]. For the monocarbide the thermal conductivity (W/m · deg) was 8.4 at 200°C, 7.9 at 300°C, and 7.9 at 400°C (0.020, 0.019, 0.019 cal/cm · sec · deg). For the dicarbide the conductivity (W/m · deg) was 23.4 at 200°C, 20.9 at 300°C, and 20.5 at 400°C (0.056, 0.050, and 0.049 cal/cm · sec · deg).

The values of the specific electrical resistivity reported by various authors [209, 213] for the monocarbide vary considerably; this may be due to the differences in composition of the specimens investigated, the properties of which were not given in full.

The specific electrical resistivity of ThC_{1-x} increases linearly with rise in temperature; its thermal emf is negative and proportional to the temperature.

The monocarbide has been obtained by heating a charge of ThO_2 + 3C in a vacuum induction furnace at 1850°C [205]. The products obtained could not be completely identified owing to the instability of ThC_{1-x} in air.

The dicarbide was first obtained when thorium ore was reduced with carbon in an attempt to obtain the metal [203]. Later this carbide was produced by heating a mixture of thorium oxide and carbon in the form of cylinders in an electric arc furnace [203, 204].

In [206] the dicarbide was obtained by reducing thorium oxide with carbon black at 2130°C. At this temperature the reaction proceeds practically to completion (99%) in 30 min.

A radiometric investigation into the removal of impurities from ThC_{2-x} during its refining has shown that at 2130°C such impurities as the isotopes of Cs, Sr, Be, Ba, Sm, Eu, Ag, and Cd are completely eliminated.

An investigation into the conditions for producing thorium carbides by reaction of the oxide with carbon black in vacuum in accordance with the equations

$$ThO_2 + 3C = ThC + 2CO,$$
$$ThO_2 + 4C = ThC_2 + 2CO,$$

made in [207], has shown that products similar in composition to ThC and ThC_2 are formed at 1800–1900°C from a charge of stoichiometric composition at a pressure of 1.33 N/m^2 ($\sim 10^{-2}$ mm Hg). The products obtained under these conditions contain, in the case of ThC_{1-x}, 4.8–9% C_{comb} and 0.1% C_{free} and, in the case of ThC_{2-x}, 9.2–9.3% C_{comb} and 0.3% C_{free}. Thus the reaction temperature of 2130°C given in [206] for producing ThC_{2-x} is too high.

Thorium carbides may also be produced by synthesis from the elements. The monocarbide ThC_{1-x} has been obtained [208] by sintering a mixture of metallic thorium and spectrographically pure carbon at 1600°C in an inert atmosphere.

Both carbides have been obtained [209] by melting a mixture of metallic thorium and spectrographically pure graphite under an argon atmosphere in an arc furnace with tungsten electrodes.

The System Uranium—Carbon. Uranium carbide was first obtained by Moissan [217], and the formula U_2C_3 was ascribed to it. Later a product was obtained whose composition corresponded more closely to the formula UC_2 [218]. Subsequent investigations confirmed the existence of the carbide UC_{2-x}, having a CaC_2-type tetragonal lattice, and the monocarbide UC_{1-x} with an NaCl-type fcc lattice [198, 219, 220].

It was concluded that the sesquicarbide is really a continuous series of solid solutions of the mono- and dicarbides. However, x-ray studies have since shown [198] that it in fact consists of a single phase with a bcc lattice. Apart from those mentioned above, no other phases have been found in the U—C system.

Data on the solubility of carbon in liquid uranium at various temperatures are given below [220]:

Temperature, °C	1350—1400	1450—1500	1550—1600	1650—1700
C, wt.%	0.10	0.17	0.36	0.42
Temperature, °C	1750—1800	1850—1900	1950—2000	2080—2130
C, wt.%	0.51	1.60	1.74	2.92

The phase diagram of the U—C system was first studied in [219] by x-ray and metallographic examinations and melting-point measurements. The investigations were extended in [214, 220-222].

The most reliable data are those given in [222], on which the diagram shown in Fig. 29 is based. According to this diagram, three carbide phases exist in the U—C system: a monocarbide UC_{1-x}, a sesquicarbide U_2C_{3-x}, and a dicarbide UC_{2-x}. The last, which crystallizes in a tetragonal lattice, undergoes a transformation into a CaF_2-type cubic lattice at temperatures above 1800°C.

The monocarbide (4.8% C) crystallizes in an NaCl-type fcc lattice, the carbon atoms being situated in the octahedral spaces in the sublattice formed by the uranium atoms.

The unit cell of UC_{1-x}, contains four molecules; the distance U—C is 0.171 nm and the distance U—U, 0.175 nm [250].

Various values are given in the literature for the lattice parameter: 0.4951 [249, 250], 0.4955 [219], and 0.4959 nm [199].

According to data in [221], the lattice parameter of UC_{1-x} increases with rise in temperature from 0.49597 nm at 18-20°C to 0.5070 nm at 1800°C.

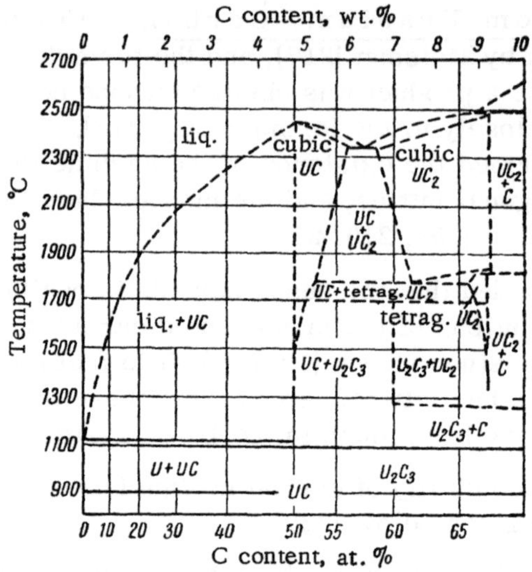

Fig. 29. Phase diagram of the system U – C [222].

The monocarbide is isomorphous with the mononitride and monoxide of uranium, and the carbon in it can be replaced by oxygen and nitrogen. Such a substitution causes a contraction in the lattice parameter of UC_{1-x}. An investigation of the system U — C — O has shown [249] that the single-phase region U(C, O) corresponds to an oxygen solubility of 12.5 at.%. The lattice parameter of the U(C, O) phase decreases with rise in oxygen content to $a = 0.4953 \pm 0.0001$ nm for the composition $UC_{0.75}O_{0.25}$. It may be assumed that the parameter value of 0.4959 nm corresponds to the carbide with limiting carbon content.

A study of the equilibrium in the system U — C — O in the temperature ranges 1360-1720 and 1500-1810°C [893] has confirmed the presence of the oxycarbide UC_xO_{1-x}, whose range of existence depends on the temperature and the ratio UO_2/C.

A neutron-diffraction study of the position of the carbon atoms in uranium carbides has shown [250] that the bond U — C is considerably weaker in U_2C_{3-x} than in UC_{1-x} and UC_{2-x} owing to the low stability and decomposition in UC_{1-x} and UC_{2-x}.

In the dicarbide, both lattice parameters increase with rise in temperature up to 1200°C; above 1200°C the value of c decreases and at 1800°C it reaches a lower value than that at room temperature.

Values of the lattice parameters of UC_{2-x} at various temperatures are given below [221]:

Temperature, °C	25	800	1200
Lattice parameter, nm:			
a	0.3516	0.3536	0.3565
c	0.5972	0 6013	0.6020
Temperature, °C	1600	1800	1900—2200
Lattice parameter, nm:			
a	0.3610	0.3660	—
c	0.5995	0.5970	0.535—0.543

This is due to the fact that at temperatures above 1800°C tetragonal UC_{2-x} is transformed into a CaF_2-type cubic lattice. The distance U — C in this phase is 0.1605 nm and the distance U — U is 0.194 nm [250]. The transformation from the tetragonal to the cubic form of UC_{2-x} has been observed in [270, 271].

The heat of transformation of the carbide of composition $UC_{1.93}$ from the tetragonal to the cubic modification at 1800°C is 46.1 kJ/kg (11.0 cal/g) [263].

Data from an x-ray investigation of U — C alloys in conjunction with chemical-analysis data show [223] that the composition of the monocarbide corresponds to the formula $UC_{0.92}$ and that of the dicarbide to $UC_{1.91}$. However, this conflicts with the phase diagram given and is not confirmed in numerous other researches.

The monocarbide UC_{1-x} is a gray powder. According to the data of various authors, its melting point varies between 2250 and 2650°C [199, 220, 227, 228].

The principal thermodynamic characteristics, which are also given in [128, 779, 951], have been corrected in accordance with the data in [251] based on an analysis of the results of a study of the U — C system.

An investigation into the behavior of the monocarbide on heating by electron bombardment in a vacuum of 133 $\mu N/m^2$ (10^{-6} mm Hg), with x-ray and electron-diffraction analyses of the condensate, has shown that UC_{1-x} evaporates without decomposing [211].

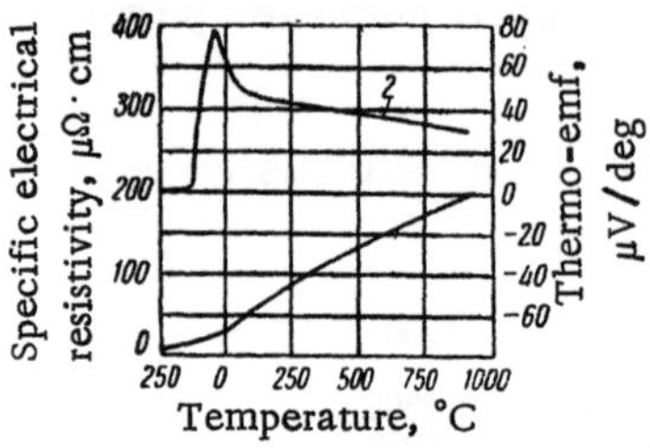

Fig. 30. Temperature dependence of the specific electrical resistivity (1) and coefficient of thermal emf (2) of UC.

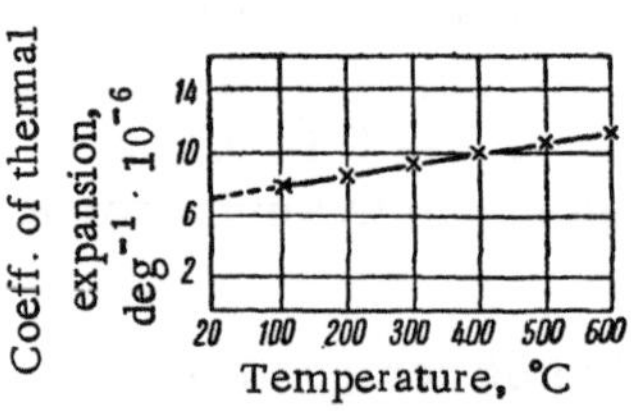

Fig. 31. Temperature dependence of the coefficient of thermal expansion of UC.

The specific electrical resistivity of UC_{1-x} is 47 μΩ · cm at room temperature [213], increasing with rise in temperature to 180 μΩ · cm at 930°C (Fig. 30); below 100°C, ρ scarcely changes with temperature. In the temperature range 1175–2050°K the specific electrical resistivity varies in accordance with the equation [253]

$$\rho = 20.4 \cdot 10^{-6} + 114.8 \cdot 10^{-9} T \ \Omega \cdot \text{cm}.$$

A study of the change in resistivity on irradiating the carbide with a dose of 10^{18} neutrons/cm^3 at 70°C has shown that ρ is increased by 2.5% as a result [254].

The thermo-emf of UC_{1-x} is positive; it reaches a maximum at 200°K (75 μV/°K), after which it slowly falls [213] (Fig. 30).

The emissive power of a dense specimen of UC_{1-x} (at 0.63 μm) decreases with rise in temperature from 0.67 at 830°C to 0.47 at 1230°C, after which it remains constant [255]. According to data in [253], its temperature dependence in the range 1150–1890°K is expressed by the equation

$$E_{0.65 \mu m} = 0.539 - 0.02T \cdot 10^{-3}.$$

The values of the thermal conductivity of the monocarbide in relation to the properties of the specimens (carbon content and porosity) are given in Table 8 [231].

According to data in [224], the thermal conductivity of UC_{1-x} at 250°C is (cal/cm · sec · deg) 0.061 at a carbon content of 4.5%, 0.055 at 4.8%, and 0.057 at 5.1% (25, 23, and 24 W/m · deg), in good agreement with the data in Table 8.

The temperature dependence of the coefficient of thermal expansion in the range 150–700°C [235] for sintered specimens of the

TABLE 8. The Thermal Conductivity of UC_{1-x} Specimens

UC		Temp., °C	Thermal conductivity	
C content, %	Density, % of theor.		$\frac{W}{m \cdot deg}$	$\frac{cal}{cm \cdot sec \cdot °C}$
5.2	—	100	25	0.060
		200	23	0.056
		300	23	0.054
		400	22	0.053
		500	23	0.054
		600	24	0.057
		700	25	0.060
		735	25	0.061
4.8	74.9	60	34	0.080
		115	31	0.074
		195	25	0.061
		265	21	0.050

monocarbide is shown in Fig. 31. A study of the coefficient of linear expansion of UC_{1-x} in the composition range 4.85-5.05% C between 100 and 2000°C [256] indicates that the coefficient is only slightly temperature dependent.

Like a number of other properties, the hardness of uranium monocarbide depends on the method by which it has been produced; it lies between 550 and 750 on the Vickers hardness scale [231].

The maximum compression strength at 25°C, 37.4 MN/m^2 (38.15 kgf/mm^2), is found in specimens produced by arc melting [231]. A study of creep in compression in the temperature range 1200–1425°C under a load of 40 MN/m^2 (4 kgf/mm^2) has shown that the creep rate at 1425°C is 0.018%/h [255].

A study of the mechanical properties of specimens whose composition was close to that calculated for the monocarbide, in the temperature range 1600–2300°C, has shown that the properties are dependent on the carbon content for the same porosity [893]. Specimens containing 4.49–4.56% C (a mixture of uranium and the monocarbide) are deformed under load more rapidly than those containing 4.72–5% C. The deformation of specimens containing 4.6–4.7% C corresponds to a change in the initial height of the specimen from 70 to 20%, while the change in specimens containing more than 4.7–5.0% C is 20–10%. This reduction in the degree of deformation is due to the presence of a liquid phase in the specimens containing 4.49–4.56% C.

Uranium monocarbide is distinguished by high stability during thermal cycling [239]. Sintered specimens with a density of 10.2 g/cm^3 (74.9% of the theoretical) have withstood 2000 cycles over the temperature range 100–750°C. According to data in [235], specimens whose density was 85% of the theoretical withstood 500 cycles over the range 200–1000°C.

A certain amount of swelling was found to occur when uranium monocarbide plates in a stainless-steel die were irradiated at 540–1000°C. The monocarbide particles had some internal porosity and reacted with the steel [231].

On irradiating UC_{1-x} specimens of various compositions obtained by arc melting in capsules filled with NaK [231], the density decreased somewhat (by 0.7–2.5%) and cracks appeared. During the process the surface of the specimen reached a maximum temperature of 580°C and the center 1000°C. The most resistant to irradiation is the uranium carbide of stoichiometric composition.

Uranium dicarbide UC_{2-x} is a gray powder. Its thermodynamic properties are given in [211, 251, 260, 261].

The specific heat of the product containing 95.3 mol.% UC_2 and 4 mol.% UC in the temperature range from −263 to +67°C is 60.75 J/mole · deg (14.50 cal/mole · deg) [128].

A study of the vapor tension of uranium over UC_{2-x}, using Knudsen's method [211] with subsequent activation analysis of the condensate, has shown that in the temperature range 2010–2893°C the heat of evaporation of uranium is 544.7 ± 21 kJ/mole (130 ± 50 kcal/mole); this is in good agreement with the data in [399], in which the contents of uranium and UC_{2-x} molecules in the gaseous phase were determined from an investigation of the evaporation of the dicarbide.

The vapor tension of uranium is represented by the equation

$$\log p_U = -\frac{27.800}{T} + 4.07,$$

and

$$\log p_{UC_{2-x}} = -\frac{40.400}{T} + 8.38.$$

The most widely used methods of producing uranium carbides are by synthesis from the elements, the reaction of uranium with carbon-containing gases, and the reduction of uranium oxides with carbon.

Until recently uranium carbides were obtained by melting a mixture of metallic uranium powder and carbon in an electric arc furnace in an argon and helium atmosphere [224].

As shown in [225], in producing uranium carbides metallic uranium powder was mixed with sugar charcoal in various proportions and, after further mixing, was moistened with paraffin diluted with benzene and compacted into briquets which were then heated in an electric resistance furnace with graphite elements in an argon atmosphere at 2300°C.

According to data in [220], the temperature at which the monocarbide is formed by synthesis from the elements is 2100°C, and that at which the dicarbide is formed 2400°C.

It has not proved possible to obtain the sesquicarbide by synthesis from the elements [227, 228]. After cooling, the reaction product consisted of a mixture of UC_{1-x} and UC_{2-x}. U_2C_{3-x} has been obtained by heating UC_{1-x} with UC_{2-x} in vacuum in the temperature range 1250–1800°C [227].

One of the most promising methods for producing uranium carbides is by the reaction of metallic uranium powder with carbon-containing gases, in particular with methane, in accordance with the equation

$$U + CH_4 = UC + 2H_2.$$

Uranium monocarbide has been obtained in this way in [219]. Hydrogen was passed over the metallic uranium powder at 225°C to form uranium hydride. On completion of hydride formation the temperature rises to 450°C and the hydride decomposes to give finely divided uranium powder. This cycle is repeated three times, at the end of which the temperature has risen to 600–900°C and methane is passed over the powder. At these temperatures uranium monocarbide is formed, while at higher temperatures the dicarbide is formed. The advantage of this method is that it leads to the formation of very finely divided carbide powders, from which it is subsequently possible to produce fairly dense compacts.

Carbide with a particle size of about 10 μm is obtained by the reaction of uranium with methane at 700°C [199].

Fischer [230] has obtained UC_{2-x} by the reaction of finely divided uranium with hydrocarbons of the methane type at temperatures somewhat below the melting point of the dicarbide.

Practically pure monocarbide can be obtained by using propane or butane at 700–750°C [216].

UO_2, U_3O_8, and UO_3 have been used as the raw materials for producing uranium carbides by the reduction of oxides with carbon.

The dioxide is reduced in accordance with the reaction

$$UO_2 + 3C = UC + 2CO,$$
$$UO_2 + 4C = UC_2 + 2CO.$$

The reaction with the formation of mono- and dicarbides has been studied in [231-235, 237-239].

Uranium carbides were first obtained by this method in [232], in which a mixture of UO_2 and sugar charcoal was heated in vacuum at 2450°C, and in [233], in which the equilibrium of the reaction $UO_2 + 4C = UC_2 + 2CO$ was investigated in the range 1480–1800°C.

A thorough investigation of the reduction of UO_2 by carbon in order to obtain the monocarbide was carried out in [235]. A manometric study of the reaction showed that with an initial pressure in the system of 0.13 N/m^2 (10^{-4} mm Hg) and a rate of temperature rise of 50 deg/min, the reaction begins at a significant rate at 1300°C.

The optimum conditions for producing UC consist in heating a mixture of UO_2 + 3C of stoichiometric composition in a vacuum at 1800°C, with a holding time after the initial pressure has been obtained in the system of 10–30 min.

In investigating the reduction of UO_2 [199, 237] by carbon in order to produce the monocarbide, it has been found that the oxide UO_2 generally used for this purpose contains somewhat more oxygen than the amount corresponding to the formula UO_2 and, moreover, that it absorbs moisture to some extent. The absorbed moisture has less effect on the process than has the oxygen content of

UO_2, which increases with time and is greater the larger the surface of the oxide powder.

If a nonstoichiometric oxide of uranium is used, the following equation must be used in determining the composition of the charge:

$$UO_{2-x} + (3 + x)\,C \rightarrow UC + (2 + x)\,CO.$$

To obtain homogeneous carbide, briquets consisting of a mixture of UO_2 and carbon black were arc melted in an argon atmosphere at a pressure of 55.65 kN/m^2 (0.5 atm). Melting takes place in two stages. In this way, 100 g of carbide can be obtained in 1 h in the form of briquets having a density of 13.58-13.60 g/cm^3 (theoretical 13.66 g/cm^3) and containing 4.75-4.88% C. To obtain large quantities of uranium carbide by melting and casting, a special apparatus is used in which a 2-kg charge can be melted with a current of 4000 A and a power of 100 kW.

When using U_3O_8 as the raw material, uranium carbides are obtained by the reactions

$$U_3O_8 + 11C = 3UC + 8CO,$$
$$U_3O_8 + 14\,C = 3UC_2 + 8CO,$$

which have been studied in [220, 225, 200]. The first reaction proceeds at 1800°C and the second at 2400°C [220, 225].

The following two reactions by which UC is formed have been investigated in [238]:

$$U_3O_8 + 9C + 2H_2 = 3UC + 6CO + 2H_2O,$$
$$U_3O_8 + 10C = 3UC + CO_2 + 6CO.$$

In the first case, the charge was heated in a stream of hydrogen at 500°C, and U_3O_8 was thereby reduced to UO_2. The reduced briquets were rapidly transferred to a vacuum furnace in which the reaction $UO_2 + 3C = UC + 2CO$ took place at 1200°C. This reaction was completed at 1430°C and then the temperature was raised to 1700°C, at which temperature the briquets were sintered for 3 h.

The products of the reaction consisted of single-phase monocarbide with a certain deficiency of carbon due to losses by the re-

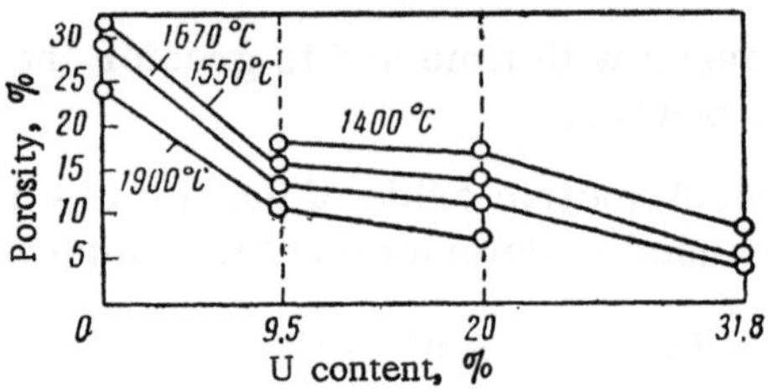

Fig. 32. Dependence of the porosity of sintered UC specimens on temperature and uranium content in the mixture.

action of carbon in the charge with water vapor formed during the reduction of U_3O_8 to UO_2. The second reaction was studied by heating the charge in vacuum. The reduction of U_3O_8 to UO_2 began at 620°C, after which the temperature was raised to 700°C and the reduction to UO_2 was completed. Then the temperature was raised to 1700°C and UO_2 was reduced to UC. In this case the carbon content was again below tha' corresponding to stoichiometric UC.

To obtain UC of stoichiometric composition the carbon content in the initial charge should exceed the calculated value by 0.5% [200].

Attempts to produce the monocarbide by using ammonium diuranate have attracted interest [917].

In [243], the conditions for producing high-density UC specimens by the following methods were investigated: by sintering UC_{1-x}, by sintering a mixture of UO_2 + C at 1500-1800°C, by compressing UC_{1-x} in a copper sheet, and by hot compaction of a mixture of uranium and graphite powders. Specimens with a porosity exceeding 25% were obtained by the first three methods.

The sintering of the monocarbide containing additions of metallic uranium in vacuum has been studied in [235]. Figure 32 shows the dependence of the porosity of the sintered specimens on the temperature and uranium content of the mixture. On sintering UC_{1-x} at 2200°C for 2 h the residual porosity is 10%. Addition of uranium lowers the porosity, and with a content of 31.8% U a porosity of 5% can be attained by sintering at 1700°C. This reduction in porosity is due to the increase in grain size of UC_{1-x} observed in the microstructure, and to the change in shape of the grains resulting from the recrystallization of the UC_{1-x} by the molten uranium.

An investigation of the conditions that affect the density of sintered specimens has shown [199, 229, 238] that slight contamination of the initial products by oxygen affects the density. At high temperatures (>1900°C), at which sintering takes place on account of the oxygen contained in the carbide, the reaction

$$2UC + UO_2 = 3U + 2CO$$

occurs, and as a result sintering proceeds in the presence of a liquid phase, which ensures that high-density specimens are obtained.

Specimens with a density up to 90% of the theoretical value have been obtained [199] by sintering UC_{1-x} produced from a mixture of metallic uranium, previously treated with a 25% solution of HNO_3, and graphite in the range 1500-2000°C. However, considerable contamination by oxygen leads to a sharp reduction in the carbon content of the final products. Thus, on sintering a product containing 5% C, 1.3% O, and 0.31% N at 1900°C, the carbon content in the specimens was 3.87%.

The density of the specimens has also been raised by making additions to the U + C mixture. Thus, a porosity in the specimens of 3.7% has been attained by adding 10% UFe_2 to the U + C mixture and sintering in vacuum at 1200-1450°C. Additions of UAl_2, UBe_{13}, and U_3Si_2 affected the sintering of the monocarbide by increasing the shrinkage and lowering the sintering temperature [252, 935].

A higher density can be attained in the specimens by hot compaction than by sintering. A study of the effect of various factors on the density of specimens produced by hot compaction has shown that specimens with a porosity of less than 5% can be obtained by this method [199, 235, 243, 276]. Specimens with such a porosity have been obtained by compacting UC_{1-x} powder containing 0.18% oxygen at 1480°C [199]. Investigation of the effect of compaction temperature and pressure has shown that the final density of the specimens depends to a considerable extent on the pressure and very little on the temperature of compaction. The authors indicate that in the hot compaction of UC_{1-x} consolidation takes place as a result of grain-boundary slip, plastic flow and volume diffusion playing a secondary part [247]. Specimens with a density of 98%

have been obtained by adding Mo, W, WC, TiC, or SiO_2 in compacting at 1000-1200°C under a pressure of 19.6 MN/m^2 (200 kgf/cm^2) [248].

Very dense specimens (98-99.5% of theoretical) containing 4.75-4.88% C have been produced by melting in an electric arc furnace under an argon or helium atmosphere [199, 231]. Tungsten, graphite, and uranium — graphite electrodes are used in the arc melting. By this method it is possible to produce specimens of any required shape and size.

Ingots of UC_{1-x}, 9-25 mm in diameter and 150 mm long, have been obtained by arc melting [272]. The density of the ingots was 99% of the theoretical value.

The System Neptunium — Carbon. The dicarbide NpC_2 has been obtained by heating NpO_2 in a graphite crucible at 2660-2800°C for 5 min [288]. A mixture of the carbides Np_2C_3 and NpC in the ratio of 1 : 5 is formed by the reaction of metallic neptunium with graphite at 1200°C.

In [936], single-phase neptunium monocarbide of composition $NpC_{0.94}$ was obtained by the reaction of carbon with neptunium hydride in a vacuum of $5 \cdot 10^{-6}$ mm Hg at 1400°C.

The sesquicarbide Np_2C_3 was obtained by the reaction of the monocarbide with carbon under the same conditions. This carbide has a cubic lattice of the U_2C_3 and Pu_2C_3 type, its parameter having a mean value between the parameters of U_2C_3 and Pu_2C_3. The authors did not succeed in obtaining the dicarbide.

The System Plutonium — Carbon. Two carbides have been identified by x-ray analysis in the system Pu — C: a monocarbide with an fcc lattice and a sesquicarbide with a bcc lattice [238]. There is also evidence of the existence of a dicarbide PuC_2, which is stable above 1750°C.

An investigation of Pu — C alloys produced by arc melting has shown [265] that the hardness of the specimens increases sharply at a carbon content above 39.7 at.%, indicating a crystallographic transformation at this composition. According to x-ray data, PuC exists at carbon contents exceeding 39.7%. From data in [236] the lower boundary of the PuC phase lies at 42.5 at.% C. At 1000°C this boundary is displaced to 45.5-46.0 at.% C. Above

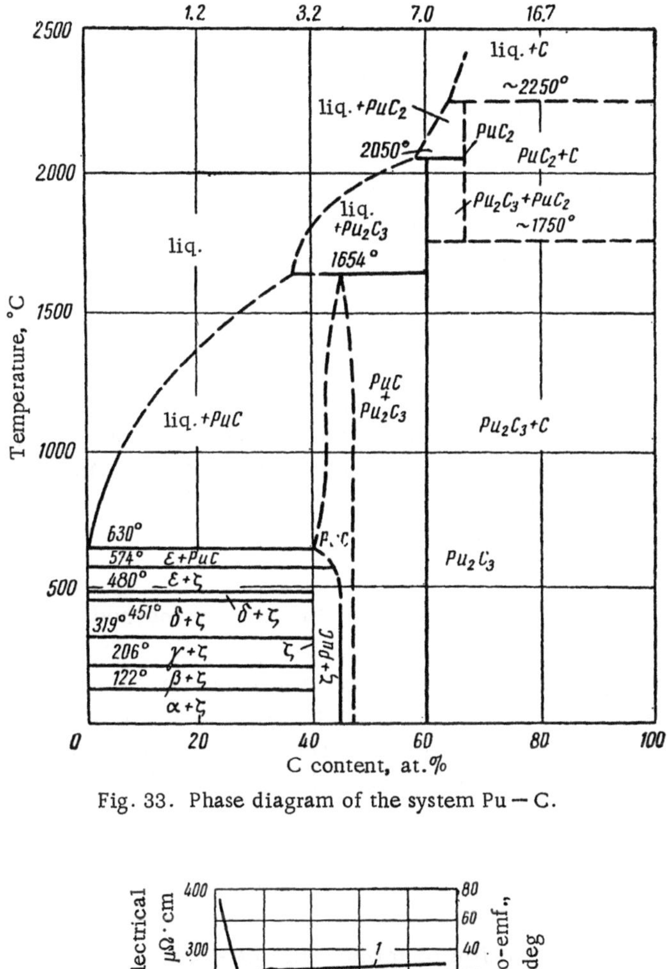

Fig. 33. Phase diagram of the system Pu – C.

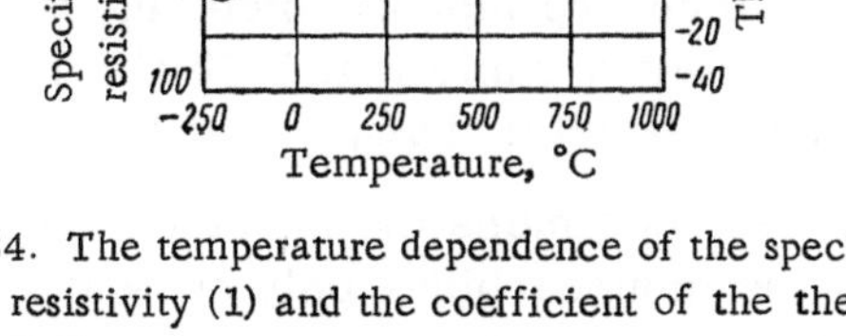

Fig. 34. The temperature dependence of the specific electrical resistivity (1) and the coefficient of the thermo-emf (2) of PuC_{1-x}.

48% C, lines corresponding to PuC and Pu_2C_3 are observed in the x-ray diffraction diagrams.

The phase diagram of the system Pu — C is shown in Fig. 33 [199]. According to this diagram, there is at 50 at.% C a mixture of PuC and Pu_2C_3, from which it follows that there is a monocarbide with a limiting content of 48 at.% C.

Investigation of Pu — C alloys containing 38-50 at.% C has confirmed [199] that the product containing 50 at.% C is two-phase and consists of PuC_{1-x} and Pu_2C_3. The stable range of the phase PuC_{1-x} corresponds to a content of 46-48% C. The ξ -phase which appears on the diagram at 40 at.% C is tentatively regarded as the compound obtained by the peritectic reaction Pu + PuC_{1-x} at 575°C.

The monocarbide has a high specific electrical resistivity which is only slightly temperature-dependent. The thermo-emf of PuC_{1-x} is positive and increases with rise in temperature (Fig. 34) [213]. Below 1730°C the monocarbide becomes antiferromagnetic. The ratio of the specific heat of PuC_{1-x} to that of UC_{1-x} is 0.2 [275].

Plutonium monocarbide is obtained by methods similar to those used for producing uranium monocarbide. Synthesis from the elements is carried out by arc melting a mixture of plutonium and graphite and also by the direct hot compaction of a mixture of plutonium and graphite at 1000°C [231]. The reaction of plutonium oxide with carbon in accordance with the equations

$$PuO_2 + (2 + x)C \rightleftarrows PuC + 2CO,$$
$$2PuO_2 + 7C = Pu_2C_3 + 4CO$$

results in the production of mono- and sesquicarbides at 1250-1450°C [199]. It has been established in [268] that the reaction of PuO_2 with carbon to form the monocarbide begins at 1100°C in vacuum and proceeds fairly rapidly at 1350°C. Two carbides are found in the reaction products: PuC_{1-x} and Pu_2C_3.

To obtain dense specimens the carbides are sintered in two stages: sintering the monocarbides only once at 1250-1450°C leads to the production of specimens with low density. After grinding and resintering in the same temperature range, the density attains 96% of theoretical for PuC_{1-x} and 99% for Pu_2C_3.

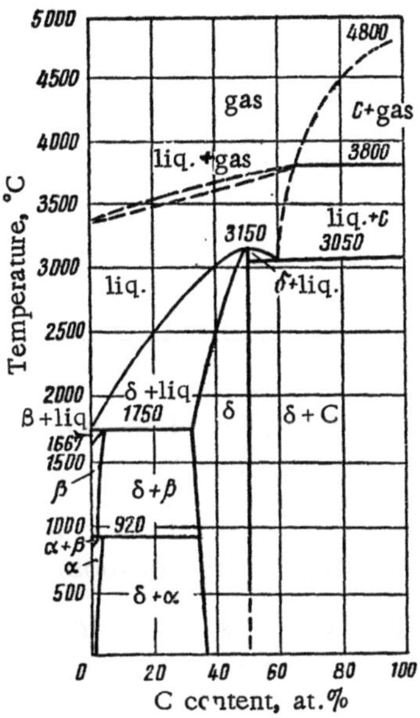

Fig. 35. Phase diagram of the system Ti — C [288].

Carbides of Metals of Subgroup IVa

The System Titanium—Carbon. The system Ti — C has been the subject of numerous researches [83, 98, 162, 168, 278-281, 284-290]. The fullest reviews of most of these researches are given in [11, 22, 277].

On the basis of these numerous investigations the phase diagram of the system Ti — C shown in Fig. 35 has been proposed [288]; according to this diagram the system contains a single carbide phase TiC_{1-x} with a wide homogeneity range. The width of this range, investigated by Cadoff and Nielsen [288] and also in [83, 98, 162, 168] over a wide range of compositions (up to 85 at.% C), extends from 37 to 50 at.% C (13-20 wt.% C). The existence of the carbides Ti_2C [282] and TiC_2 [291] has not been confirmed.

Carbon raises the $\alpha \rightleftharpoons \beta$ titanium transformation temperature from 882 to 920°C [288-290]. At 920°C the peritectic transformation $\beta + TiC \rightleftharpoons \alpha$ occurs. At this temperature the solubility of carbon in α-titanium is a maximum, and it falls with reduction

TABLE 9. Evaporation Rate of Titanium Carbide and Vapor Pressure of Titanium and Carbon

Temperature, °C	Evaporation rate, $g/cm^2 \cdot sec \cdot 10^{-7}$	Ti + C vapor pressure	
		$\mu N/m^2$	mm Hg
1800	1.35	2.666	$0.020 \cdot 10^{-6}$
1900	4.38	8.931	$0.067 \cdot 10^{-6}$
2000	7.78	15.99	$0.120 \cdot 10^{-6}$

in temperature. The solubility of carbon in β-titanium is a maximum at 1750°C and also falls with reduction in temperature.

The lattice parameter of titanium carbide varies between 0.430 and 0.433 nm, depending on its composition.

Data on the dependence of the lattice parameter, in nm, of TiC_{1-x} on carbon content are given below [327, 864]:

Composition	Parameter	Ref.	Composition	Parameter	Ref.
$TiC_{0.33}$	0.4313	[327]	$TiC_{0.5}$	0.42990 ± 10^{-4}	[864]
$TiC_{0.85}$	0.4315*	[327]	$TiC_{0.6}$	$0.43120 \pm 5 \cdot 10^{-5}$	[864]
$TiC_{0.92}$	0.4325	[327]	$TiC_{0.7}$	0.43215 ± 10^{-4}	[864]
$TiC_{0.96}$	0.4329	[327]	$TiC_{0.8}$	$0.43250 \pm 5 \cdot 10^{-5}$	[864]
$TiC_{0.96}$	0.4320	[327]	$TiC_{0.9}$	$0.43290 \pm 5 \cdot 10^{-5}$	[864]

* This specimen contains 1.53% N and 0.86% O.

For TiC of stoichiometric composition the parameter is 0.4315–0.4324 nm [286, 315, 328–333].

The differences in the values of the parameter are due to the fact that the carbide is always somewhat contaminated with oxygen; as was shown in [334], this causes a contraction of the lattice and a decrease in its parameter.

A high-temperature x-ray study of the carbide of composition $TiC_{0.921}$ has revealed [335] the following temperature dependence of the lattice parameter and coefficient of linear expansion:

Temperature, °C	a, nm	α_{298}^{T} $deg^{-1} \cdot 10^{-6}$
23	0.43178± ±0.00003	—
500	0.43322± ±0.00004	6.99±0.34

Temperature, °C	a, nm	α_{298}^{T} deg^{-1} · 10^{-6}
702	0.43395± ±0.00004	6.99 ± 0.34
848	0.43449± ±0.00004	7.61±0.20

An x-ray study of the electron density distribution in titanium carbide has shown [240, 245] that the titanium atoms in the carbide have a positive charge + 4, while the carbon atoms have a certain negative charge. This indicates that the ionic bond makes a definite contribution to the complex bond between the titanium and carbon atoms.

A mass-spectrographic investigation has shown that, on heating in vacuum, titanium carbide dissociates into gaseous atomic titanium and carbon [336].

The evaporation rate and vapor pressure of titanium and carbon in relation to temperature are given in Table 9 [337].

The thermodynamic properties of titanium carbide have been studied in [337-342, 673].

The dependence of the enthalpy of the carbides on their carbon content has been investigated in [342]. Specimens were prepared by sintering compressed mixtures of hydrogenated iodide titanium and graphite in vacuum at 1800-2000°C for 3 h. Analysis of the products obtained indicated the formation of oxycarbides.

The dependence of the enthalpy in the composition range $TiC_{0.79}$ — $TiC_{1.0}$ on the carbon content is linear. The enthalpy of oxycarbide formation is higher than that corresponding to the extrapolated values for carbides with the same carbon contents; thus, the enthalpy of titanium oxycarbide formation is the sum of the enthalpies of carbon and oxygen separately combined with titanium.

The heat of formation of titanium carbide from its elements has been determined in [336, 338, 339, 673]. The data given by the various authors differ somewhat owing to the difference in composition of the specimens on which the determinations were carried out. The most reliable data are those given in [336, 339, 673], which agree well among themselves and give for the carbide, whose composition is close to the calculated one, a value of ΔH_{298} = −183.7 ± 1.63 kJ/mole (−43.85 ± 0.39 kcal/mole).

The thermal conductivity of TiC_{1-x} in the temperature range 500-2100°C is 37.7 W/m · deg (0.09 cal/cm · sec · deg) at 600°C and 46 W/m · deg (0.11 cal/cm · sec · deg) at 2000°C [343].

The thermal conductivities of specimens obtained by various investigators have been determined in [344], and there is some variation in the results. Thus, at 965-975°C the thermal conductivities of the specimens were 41 and 30 W/m · deg (0.098 and 0.073 cal/cm · sec · deg) and at 1415-1425°C the values were 44 and 31 W/m · deg (0.0105 and 0.075 cal/cm · sec · deg); this variation is evidently due to the difference in composition of the specimens. Measurements of the thermal conductivity of $TiC_{0.97}$ at −183 and 0°C gave values of 34 and 80 W/m · deg (0.082 and 0.19 cal/cm · sec · deg), respectively [349].

The specific electrical resistivity of titanium carbide increases linearly with rise in temperature [345, 346].

The temperature dependence of the specific electrical resistivity and of the Hall effect of the compound $TiC_{0.96}$ has been investigated in [343]. The results indicate an increase in the specific electrical resistivity and Hall coefficient with rise in temperature, thus confirming the metallic nature of the conduction in titanium carbide.

A study of the temperature dependence of the specific electrical resistivity and of the thermo-emf coefficient for titanium carbide specimens of several different compositions within the homogeneity range [347], and also of the Hall coefficient and the magnetic susceptibility at room temperature, shows that the specific resistivity decreases with rise in carbon content, while the thermo-emf coefficients increase. A similar variation in the specific resistivity has been observed in [864]. However, it was shown in [703, 715] that such a variation in the resistivity occurs in specimens having a low degree of purity; if the titanium carbides are obtained from very pure components in a high vacuum, the reverse relationship exists. The variation in the Hall coefficient is similar to that of the thermo-emf coefficients and agrees with them in sign (Table 10).

The negative values of the Hall and thermo-emf coefficients indicate the predominantly electron character of the electrical conduction of titanium carbide in the homogeneity range.

TABLE 10. Some Physical Properties of TiC_{1-x} in the Homogeneity Range at Room Temperature

Formula of compound	Spec. electrical resistivity, μΩ·cm	Thermo-emf coeff., μV/deg	Hall coeff., R 10^4, cm/C
$TiC_{0.59}$	174	−7.7 ±0.2	−4.0
$TiC_{0.72}$	131	−8.8 ±0.3	−4.85
$TiC_{0.81}$	104	−10.4±0.3	−6.5
$TiC_{0.988}$	52.5	−12.5±0.2	−6.7

An investigation of the temperature dependence of the electrical resistivity and thermo-emf, carried out on single crystals, has shown [348] that the resistivity increases with rise in temperature; this also points to the metallic type of conduction in TiC. The resistivity at room temperature is 200 ± 2 μΩ · cm; this is in good agreement with the data in [397], according to which the specific resistivity of a specimen of pyrolytic $TiC_{0.92}$, very similar to a single crystal, is 200 μΩ · cm.

Measurements of the microhardness of titanium carbide in the homogeneity range has shown [333] that the hardness varies linearly and on extrapolation to zero carbon content the microhardness corresponds to the value for pure titanium.

The mechanical properties of titanium carbide specimens have been studied in [350–352, 919]. They are found to depend on porosity. On changing from dense specimens to specimens with 5% porosity, the strength of titanium carbide in bending decreases by almost half. It increases considerably with reduction in the particle size of the original carbide.

In [350] a study was made of the dependence of the mechanical properties and electrical resistivity of titanium carbide specimens obtained by hot compaction at 3000°C for 30 sec, on the coarseness of the original carbide grains. It was found that the mechanical strength increases with reduction in the grain size of the original powder, while the electrical resistivity decreases.

The temperature dependence of the ultimate strength of TiC_{1-x} specimens in tension and compression is shown in Table 11 [152, 351]. The porosity of the specimens varied from 1 to 20%.

TABLE 11. Ultimate Strength of TiC_{1-x} Specimens in Tension and Compression (Porosity of Specimens = 1-20%).

Temperature, °C	σ_{ten} MN/m²	σ_{ten} kgf/mm²	σ_{comp} MN/m²	σ_{comp} kgf/mm²
20	560—1050	56—105	1380	138
800	380	38.0	—	—
1000	280	28.0	875	87.5
1200	—	—	521	52.1
1300	250	0.25	—	—
1400	—	—	350	35.0
1600	—	—	230	23.0
1800	—	—	310	31.0
2000	—	—	164	16.4
2200	—	—	94.5	9.45

A study of the effect of neutron irradiation on the properties of titanium carbide has shown [267] that it gives rise to an increase in the electrical resistivity, microhardness, and brittleness and to a decrease in the grain size; the original properties are restored by heat treating the irradiated specimens.

Moissan [246, 292] obtained a carbide corresponding to the formula TiC by the electric arc melting of a charge consisting of titanium dioxide and carbon in the calculated proportions.

The method of producing titanium carbide by reducing the oxide with carbon black at lower temperatures (before the charge melts) has been used with slight variations in many researches [25, 293-295] and is used for the commercial production of the carbide. However, the reaction products usually contain an appreciable amount of free carbon; some reduction in the amount can be obtained by using special methods.

Thus, on heating titanium carbide containing free carbon with manganese, the carbon is combined in the carbide Mn_3C, which is then removed by solution in dilute hydrochloric acid. Some authors have used an addition of up to 1% chromium oxide for the same purpose [296]. Another method that has been used is to separate the free carbon by flotation after mixing the finely ground product with an 0.1-1.5% soap solution.

It has been possible to raise the combined carbon content to 18.6% by treating titanium carbide containing 17.3% C_{comb} with a mixture of hydrogen and methane with chloroform vapor [298].

It was also found in [298] that the rate of formation of carbide and the completeness of the reaction are increased by introducing chlorine-containing compounds into the furnace atmosphere. The compound containing chlorine, for example CCl_4, chlorinates the titanium oxide, and the titanium chloride thus formed reacts with hydrogen and carbon in accordance with the equation

$$TiCl_4 + 2H_2 + C = TiC + 4HCl.$$

An investigation of this reaction has shown that it proceeds at a significant rate at temperatures above 1100°C [299].

The reaction rate increases if the hydrogen is replaced by iron:

$$TiCl_4 + 2Fe + C = TiC + 2FeCl_2,$$

and the reaction proceeds at lower temperatures.

The effect of additions of hydrogen-reduced iron, alumina, quartz sand, calcium oxide, and metallic silicon on the reduction of titanium dioxide by carbon has been investigated in [300].

It was found that the iron considerably intensifies the reduction of titanium oxide and displaces it in the lower temperature direction.

The addition of Al_2O_3 to the charge similarly improves the conditions for reducing TiO_2 and also displaces the process in the lower temperature direction.

Additions of CaO, SiO_2, and metallic silicon have no appreciable effect on the process.

The mechanism of titanium carbide production by reducing the dioxide with carbon has been very thoroughly studied in [45-47, 302, 310, 311]. It was found that the process of reduction by carbon passes through a number of stages in which the lower oxides Ti_3O_5, Ti_2O_3, and TiO are formed, the last stage consisting of the reduction of TiO:

$$TiO + C = TiC + CO.$$

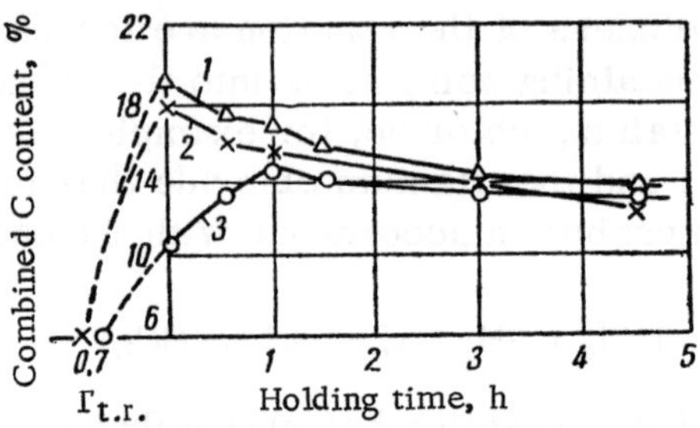

Fig. 36. Dependence of the C_{comb} content in the products of reducing TiO_2 with carbon on the holding time at 1900°C in a hydrogen atmosphere: 1) charge of TiO_2 + lamp black; 2) charge of TiO_2 + carbon black; 3) charge of TiO_2 + sugar carbon. $T_{t.r.}$) Time of initial temperature rise.

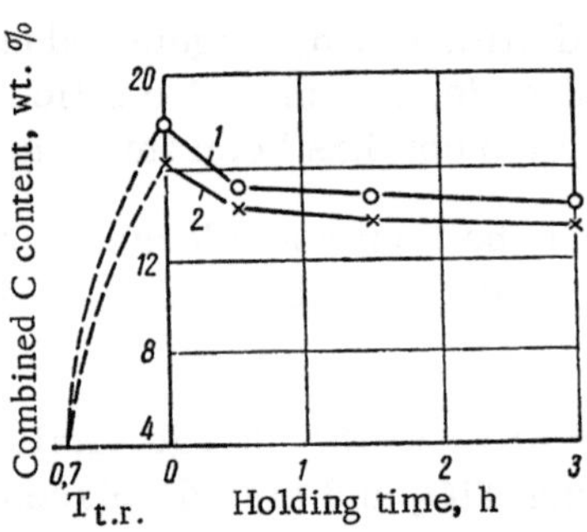

Fig. 37. Dependence of C_{comb} content in products of reduction of TiO_2 with carbon on holding time at 1900°C in a CO atmosphere: 1) charge of TiO_2 + lamp black; 2) charge of TiO_2 + carbon black.

Figure 36 shows the dependence of the combined carbon content in the products of the reaction of TiO_2 with various reducing agents on the holding time at 1900°C in a hydrogen atmosphere. It can be seen from the figure that the most active reducing agent for titanium oxide is lamp black.

Heating a mixture of TiO_2 with lamp black and carbon black leads to maximum combined carbon content at the beginning of the process and to its decrease with increasing holding time (Fig. 37).

The rate of temperature rise also considerably affects the process of titanium carbide formation. This is explained by the fact that heating the charge at a high temperature leads to an increase in the size of the carbon black crystals, i.e., to its graphitization. As a result, the concentration of acetylene in the furnace atmosphere decreases and falls below the equilibrium value corresponding to maximum combined carbon content in the solid phase. On this account the combined carbon content decreases, i.e., there is decarbidization of the titanium carbide. The resulting reaction products consist of the solid solution Ti(CO) or Ti(CNO).

A study of the effect of CO pressure on the composition of the products of the reduction of TiO_2 with carbon in vacuum has shown [307, 308] that titanium carbide saturated to the maximum extent with carbon is formed at a pressure of 133.3-1333 N/m^2 (1-10 mm Hg).

In discussing the final stage of the reduction TiO + C = TiC + CO, Meerson [312] indicates that as the concentration of oxygen in titanium approaches 0.05% the thermodynamic potential for the formation of the Ti — O bond is comparable to that of the bond Ca — O. As a result the thermodynamic potential for the reaction Ti(O) + 2C = TiC + CO has been calculated and so has the value $K_p = P_{CO}$ for 1300°C, which was 130.6 N/m^2 (9.8 mm Hg); this is in good agreement with the experimental data given above [307]. At higher temperatures there is a discrepancy between the calculated and the experimental values of the equilibrium CO pressures, which increases with rise in temperature, the calculated values becoming considerably higher than those obtained experimentally. Consequently, as the temperature rises, the Ti — O bond energy in the Ti — C — O solid solution becomes greater than that of the Ca — O bond.

According to [313], the optimum temperature for TiC formation is 1800°C; with slow heating the reaction proceeds in the stages $TiO_2 \rightarrow TiO \rightarrow Ti \rightarrow TiC$, while with rapid heating the stages are $TiO_2 \rightarrow TiO \rightarrow TiC$.

Pure titanium carbide containing 20.0% C_{comb} was obtained in [892] by heating a mixture of pure titanium powder and acetylene soot containing 7.5-10% titanium in excess of the calculated proportions, at 1747°C and in a vacuum of 266.6 $\mu N/m^2$ ($2 \cdot 10^{-5}$ mm Hg); this excess was evidently necessary on account of some volatility of the titanium.

TABLE 12. Free-Energy Changes in Possible Reactions in the Formation of TiC_{1-x} by Deposition from the Gaseous Phase [318]

Reaction	ΔF^0_T at temperature, °C							
	900		1050		1300		1600	
	kJ/mole	kcal/mole	kJ/mole	kcal/mole	kJ/mole	kcal/mole	kJ/mole	kcal/mole
$TiCl_{4(gas)}+CH_4=TiC+4HCl$	+9.6	+2.3	−20.5	−4.9	−75.0	−17.9	−129.5	−30.9
$TiCl_{4(gas)}+2H_2=Ti+4HCl$	+210.7	+50.3	+192.3	+45.9	+158.0	+37.7	+120.7	+28.8
$TiCl_{4(gas)}+2H_2+C=TiC+4HCl$	+39.4	+9.4	+23.0	+[illegible].5	−7.9	−1.9	−34.8	−8.3
$TiCl_{4(gas)}+2C=CCl_4+TiC$	+495.2	+118.2	+502.4	+119.9	+507	+121.0	+521	+124.3
$4TiCl_{3(gas)}+7C=3CCl_4+4TiC$	+1390	+331.8	+1404	+335.2	+1452	+346.6	+1536	+366.6
$2TiCl_{2(cond)}+3C=CCl_4+2TiC$	+399	+95.3	+395.5	+94.4	+387.1	+92.4	+386.3	+92 2
$TiCl_{4(gas)}+\frac{1}{2}H_2=TiCl_{3(gas)}+HCl$	+40.2	+9.6	+29.3	+7.0	+12.6	+3.0	−5.9	−1.4
$TiCl_{3(gas)}+\frac{1}{2}H_2=TiCl_{2(cond)}+HCl$	+30.2	+7.2	+39.4	+9.4	+47.8	+11.4	+60.3	+14.4
$TiCl_{2(cond)}+H_2=Ti+2HCl$	+140.4	+33.5	+123.6	+29.5	+97.6	+23.3	+66.2	+15.8
$3TiCl_{2(cond)}=Ti+2TiCl_{3(gas)}$	+80.0	+19.1	+44.8	+10.7	+2.1	+0.5	−54.5	−13.0
$TiCl_{2(cond)}+H_2+C=TiC+2HCl$	−31.0	−7.4	−45.7	−10.9	−68.3	−16.3	−91.8	−21.9
$3TiCl_{2(cond)}+C=TiC+2TiCl_{3(gas)}$	−96.8	−23.1	−107.7	−25.7	−145.4	−34.7	−182.7	−43.6

The purest titanium carbide is obtained by deposition from the gaseous phase. In [314] it was deposited on a tungsten wire at 1300–1700°C from a mixture of titanium tetrachloride vapor, hydrogen, and toluene; according to [287], the optimum temperature for this process is 1600–2000°C.

Titanium carbide has also been deposited from a mixture of $TiCl_4$ vapor and hydrogen on a carbon filament at 1530–1830°C [315]. The product contained some excess titanium and corresponded to the formula $Ti_{1.1}C$. After annealing in a high vacuum at 1900–2100°C, the excess dissolved titanium evaporated, and a carbide containing 19.6% C was obtained.

In [316] titanium carbide was obtained by deposition on a tungsten wire at 1300–1700°C from a mixture of $TiCl_4$ vapor, hydrogen, and hydrocarbons.

The carbide has also been produced by passing a mixture of hydrogen and methane in the proportions 15 : 1, saturated with $TiCl_4$ vapor, over a tungsten filament at 1500°C [317].

It has been shown in numerous investigations that titanium carbide is deposited from the gaseous phase only in the presence of hydrogen. Thermodynamic calculations relating to the reactions possible in the formation of titanium carbide by the method described were made in [318] with the object of clarifying the part played by hydrogen (Table 12). It was shown that in the reduction of $TiCl_4$ by hydrogen, metallic titanium is precipitated which reacts with the carbon formed by the thermal decomposition of methane.

The deposition of titanium carbide from the gaseous phase has been thoroughly investigated [319], using the apparatus shown in Fig. 38. The results show that the reduction of $TiCl_4$ vapor to titanium by hydrogen scarcely takes place at temperatures up to 1750°C. In the presence of carbon the reaction proceeds with the formation of titanium carbide, the role of the carbon being to bind the titanium.

The precipitation of carbon from the gaseous mixture of $TiCl_4$, CH_4, and H_2 is due to the decomposition of the methane at the surface of the heated filament. The rate of deposition of titanium carbide increases linearly with rise in the partial pressure of CH_4 up to 1.4 kN/m^2 ($1.4 \cdot 10^{-2}$ atm), after which the increase is negligible (Fig. 39). This arises from the fact that at low partial

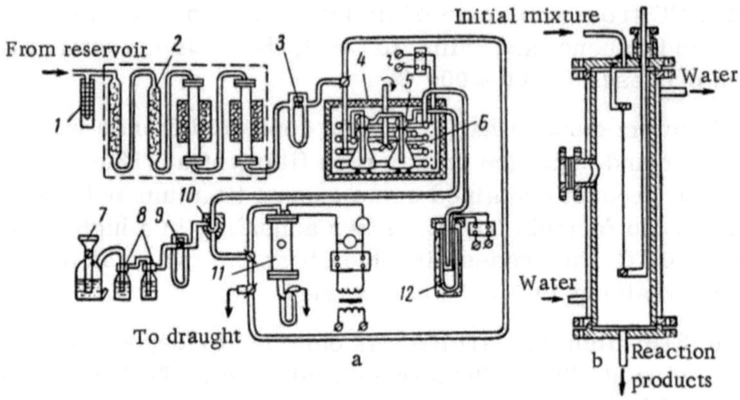

Fig. 38. Apparatus for investigating the production of titanium carbide by deposition from the gaseous phase. a) Diagram of the apparatus: 1) monostat; 2) purification system; 3) flowmeter; 4) thermostat; 5) evaporator; 6) coil; 7) gasholder; 8) flask containing H_2SO_4; 9) flowmeter; 10) Venturi nozzle; 11) reaction chamber; 12) $TiCl_4$ evaporator. b) Reaction chamber.

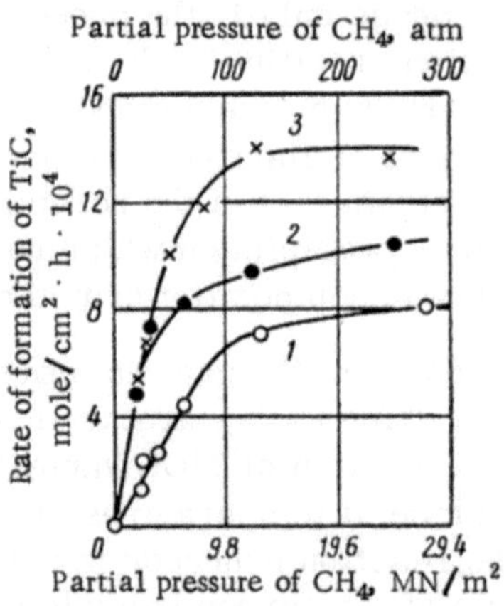

Fig. 39. Dependence of the rate of deposition of titanium carbide on partial pressure of CH_4 in mixtures, at temperatures of (°C): 1) 1320; 2) 1680; 3) 1900.

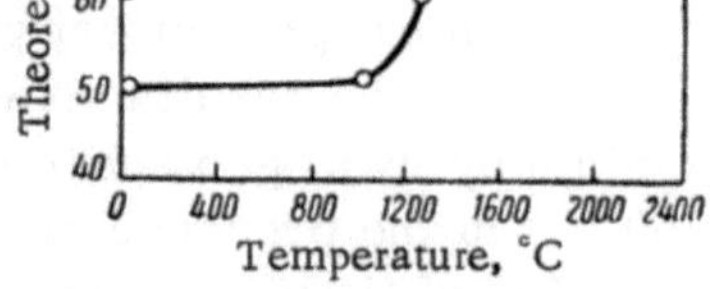

Fig. 40. Dependence of density of titanium carbide specimens on sintering temperature.

pressures of CH_4 the limiting process is the decomposition of methane, while at higher partial pressures the limiting factor becomes the reaction of titanium with carbon.

A method has been developed for producing titanium carbide by the reaction of metallic titanium or titanium oxide powder with a mixture of CCl_4 and $CHCl_3$ vapors. In addition to providing a source of carbon, the vapors act catalytically [320].

Instead of the dioxide, titanium nitride and sulfide [322] can be used as the raw material that is reduced by carbon.

Titanium carbide has also been obtained by heating an intimate mixture of titanium phosphate and carbon in an argon atmosphere at 1100-1600°C, or a mixture of titanium hydrate, finely divided carbon, and phosphoric acid.

The most efficient method of producing titanium carbide, and the one most widely used in industry, is by the reduction of titanium dioxide by carbon black in carbon-resistance furnaces in a hydrogen atmosphere or in three-phase electrode furnaces in a CO atmosphere formed during the reaction at 2300-2700°C.

Dense, compact specimens of titanium carbide containing 18.9% C_{comb} and 1.2% C_{free} have been obtained from powder of grain size 1 μm previously treated with dilute sulfuric acid [324]. After this treatment the powder is dried, compacted at a pressure of 68.6 MN/m^2 (70 kgf/mm^2) and sintered for 1 h in a vacuum of 133.3 N/m^2–533.2 mN/m^2 ($1\text{–}4 \cdot 10^{-4}$ mm Hg). The relationship between the density of the specimens and the sintering temperature is given in Fig. 40. The residual porosity was 5% with a sintering temperature of 1800°C and 3% with 2050°C.

In studying the effect of preliminary cleaning and surface activation of the titanium carbide particles [325], carbide powder containing 18.1 C_{comb} and 0.3% C_{free} was washed with a mixture of hydrofluoric and nitric acids to an extent corresponding to the solution of 2% of the total amount of carbide. Then compacted briquets of this powder were sintered at 2600°C (Fig. 41a) and 2900-2950°C (Fig. 41b). In this way it was possible to obtain specimens with a residual porosity of 4% by sintering for 7-10 min at 2900°C (Fig. 41). However, it is rather difficult in practice to carry out sintering at such high temperatures, which are very close to the melting point of titanium carbide.

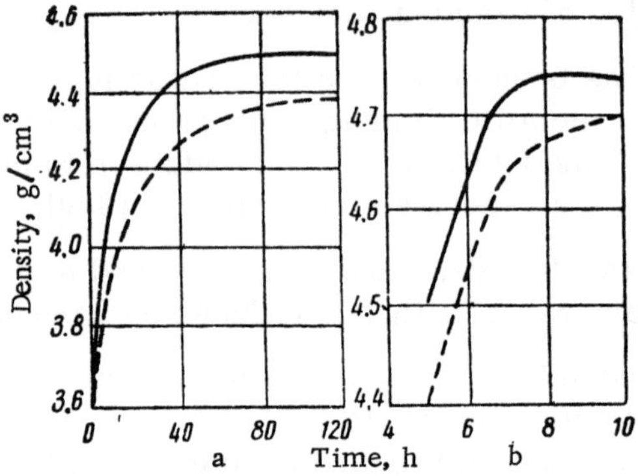

Fig. 41. Dependence of the density of sintered TiC specimens on sintering time and preliminary treatment of the powder. Continuous lines) previously washed powder; broken lines) unwashed powder. Sintering temperature: a) 2600°C; b) 2900-2950°C.

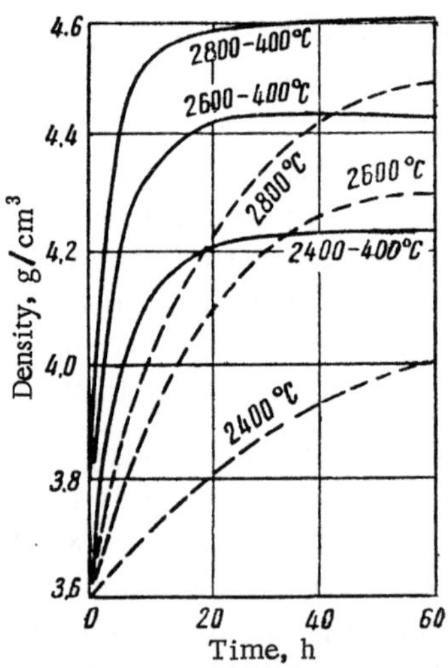

Fig. 42. Dependence of the density of TiC specimens on sintering time and temperature in cyclic (continuous lines) and isothermal (broken lines) sintering.

TABLE 13. Some Characteristics of the Range of Existence of ZrC_{1-x} as Given by Various Authors

Author and reference	C content (lower phase boundary)		Lattice param., nm	Limiting $C_{comb.}$ content		C content (upper phase boundary)		Lattice param., nm
	wt. %	at. %		wt. %	at. %	wt. %	at. %	
Umanskii [372]	4.8	27.6	0.4656	10.6	47.4	11.62	50	0.4683
Koval'skii and Makarenko [373]	7.08	36.6	0.4652	10.2	46.3	11.62	50	—
Samsonov and Rozinova [374]	3.5	21.6	0.4582	10.3	45.8	11.62	50	0.4583
Benesowsky and Rudy [375]	6 63	35.0	0.4675	—	—	11.62	50	0.4694
Sara [859]	7.6	38.5	—	—	—	11.62	50	—

An investigation into the effect of cyclic sintering on the density of the specimens [325] has shown that such sintering has a considerable effect (Fig. 42). The cyclic sintering conditions consisted of heating at the maximum temperature for 3 min, cooling to 400°C, and reheating.

The kinetics of sintering porous compacts made from carbide activated by the addition of cobalt chloride has been studied in [266], and it was shown that the addition activates the process considerably.

The System Zirconium—Carbon. According to the phase diagram proposed in [371], there is only one carbide ZrC (11.64 wt.%) in the Zr — ZrC system. This system has been further studied in [372-375]. As a result of all the investigations, Benesowsky and Rudy [375] have put forward a hypothetical phase diagram in which a single carbide ZrC_{1-x} exists. The data of the various authors on the range of existence of the monocarbide differ as regards the lower boundary (Table 13). The most reliable data are those in [859], which are close to those in [373].

The phase diagram of the Zr — C system shown in Fig. 43 was determined by x-ray, metallographic, and differential thermal analyses [859]. This diagram confirms the existence of a single

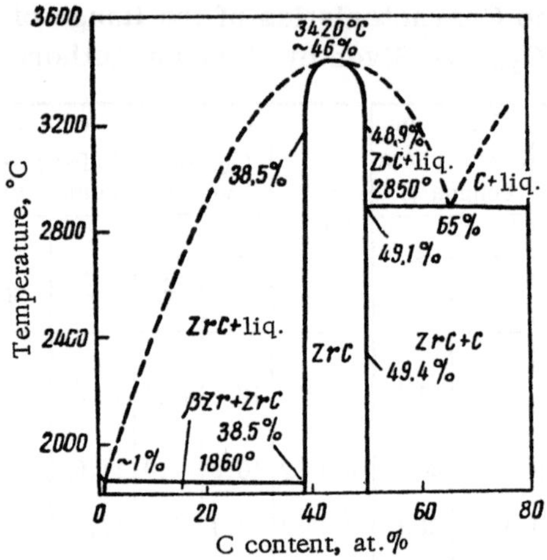

Fig. 43. Phase diagram of the system Zr – C [859].

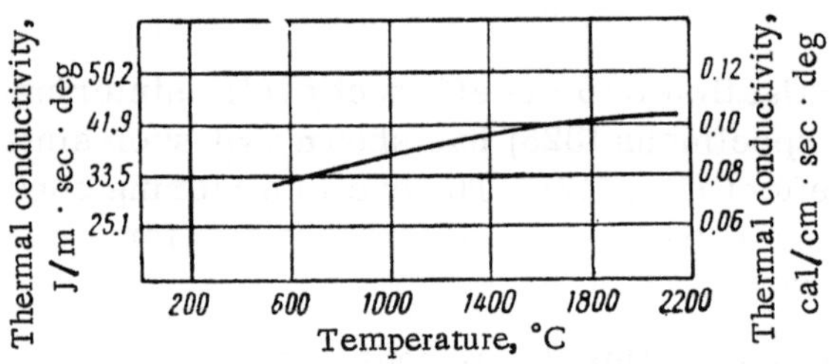

Fig. 44. Temperature dependence of the thermal conductivity of ZrC.

phase ZrC_{1-x}, lying within the composition range 38.5–50 at.% C. It crystallized in a NaCl-type fcc lattice.

This is in good agreement with the data in [928], according to which the homogeneity range of zirconium carbide extends from 34.4 to 50.0 at.% C ($ZrC_{0.62}$ – $ZrC_{1.00}$). The lattice parameters in the homogeneity range are 0.46934 nm for $ZrC_{0.62}$ and 0.46983 nm for $ZrC_{1.00}$.

Zirconium carbide forms a eutectic with carbon at 65 at.% C, 2850°C [859]; this is in good agreement with the data in [289, 716].

The values of the lattice parameter of the carbide given by numerous authors [329, 383–385] vary on account of the difference in composition of the specimens investigated.

TABLE 14. Dependence of the Lattice Parameter and Coefficient of Linear Expansion of ZrC_{1-x} on Composition and Temperature [335, 392]

Phase	a (nm) at temperature, °C					$\alpha \cdot 10^{-6}$ deg^{-1} at temperature, °C			
	−190	23	498	702	994	(−190) ÷ (÷26)	498	702	994
$ZrC_{0.80}$	0.46961± ±0.00004	0.46980± ±0.00004	—	—	—	1.9±0.8	—	—	—
$ZrC_{0.95}$	0.46957± ±0.00006	0.47025± ±0.00002	—	—	—	3.7±0.6	—	—	—
$ZrC_{0.957}$	—	0.46828± ±0.00003	0.46978± ±0.00004	0.47046± ±0.00004	0.47146± ±0.00004	—	6.73± ±0.32	6.86± ±0.22	6.99± ±0.16
$ZrC_{0.97}$	0.46951± ±0.00001	0.46986± ±0.00004	—	—	—	3.5±0.4	—	—	—

The results of an x-ray study of zirconium carbides of various compositions are given for a wide range of temperatures in Table 14 [355, 392].

The microhardness of ZrC in the homogeneity range increases linearly with rise in carbon content [373, 374, 386]. The microhardness values for the carbide of limiting composition are, according to various authors, 27,000-29,000 MN/m^2 (2700-2900 kgf/mm^2).

According to [388], the temperature dependence of the thermal conductivity of the carbide $ZrC_{0.93}$ over the temperature range 530-2100°C is represented by the curve shown in Fig. 44.

The thermoelectric properties of the carbide have been studied in [387]. The temperature dependence of the electrical resistivity and thermo-emf indicates that the carbide has electron conduction.

The specific electrical resistivity, coefficient of thermo-emf, thermal conductivity, and Hall coefficient have been investigated in the homogeneity range from $ZrC_{0.58}$ to $ZrC_{0.96}$ [391, 703]. It was shown that the addition of 1.4% N to the specimens has a marked effect on the properties of zirconium carbide. In the case of carbides synthesized from pure zirconium and carbon black in high vacuum, which results in a nitrogen content up to 0.09%, the value of the specific electrical resistivity falls with decrease in the carbon content. The thermo-emf and the Hall coefficient behave similarly, while the thermal conductivity increases.

The equation for the temperature dependence of the specific heat derived from the calculation of the specific heat by Debye's method in the temperature range 25-2730°C is given in [542, 844].

From an investigation of the evaporation of zirconium carbide [337, 389], the heat of formation of ZrC_{1-x} from the elements has been calculated; it is -199.4 ± 12.6 J/mole (-47.6 ± 3 cal/mole) for a product close to ZrC in composition. In [540], values for the heat content of $ZrC_{0.98}$ in the range 1300-2500°K have been determined by the method of mixing, and the temperature dependence of the enthalpy has been derived; this is represented by the equation

$$\Delta H_{298}^{T} = 6.018\,T + 2.192 \cdot 10^{-3}\,T^2 - 381.0.$$

A molten product whose composition corresponded to the carbide ZrC_{1-x} was first obtained by reducing zirconium ore with carbon black in an electric arc furnace [11]. Zirconium carbide has also been obtained in the arc furnace, irrespective of the content of carbon and zirconium dioxide in the charge; on cooling the melt, excess carbon is precipitated in the form of graphite [377]. Another method of obtaining the carbide is by heating a mixture of the dioxide and carbon black in a hydrogen atmosphere at 1900°C [25].

According to [22], zirconium carbide can also be obtained by heating a charge consisting of 78.75% dioxide and 21.25% carbonized sugar in a graphite-resistance furnace at 2400°C. The reaction products contain 11.3% combined carbon, 88.92% zirconium, and traces of free carbon. However, the products obtained by reducing the dioxide with carbon do not usually correspond to the composition ZrC; as a rule they contain a certain amount of oxygen and nitrogen owing to the formation of stable ZrC — ZrO — ZrN solid solutions.

Commercially pure zirconium carbide is produced on an industrial scale by melting zirconium sand containing 67% ZrO_2 with carbon in an arc furnace [378]. Scrap graphite (from crucibles and electrodes) is used as the reducing agent.

The reduction of $CaO \cdot ZrO_2$ by metallurgical or oil coke taken in small excess over the stoichiometric requirement, by heating at 1980-2200°C for 1 h, leads to the formation of a sintered cake which is treated with hot water and acid. The insoluble residue consists of practically pure zirconium carbide contaminated with 0.2% Ca.

It is desirable to carry out the reduction in vacuum so as to exclude the possibility of nitriding; in addition, by displacing the equilibrium of the reaction, the method enables the reaction to be carried out at lower temperatures.

Equilibrium in the system ZrO_2 — C was first investigated in vacuum in the temperature range 1530-1740°C [380] on the assumption that the reaction proceeds according to the equation

$$ZrO_2 + 3C = ZrC + 2CO.$$

However, owing to the lack of data on the complete chemical and x-ray analyses of the reaction products, there is no sound basis for this assumption. An investigation into the mechanism of the reduction of zirconium dioxide by carbon in a vacuum of 0.66-53.2 N/m^2 ($5 \cdot 10^{-4}$-$4 \cdot 10^{-1}$ mm Hg) in the temperature range 800-1950°C has shown [21] that the reduction and carburization of the dioxide in vacuum proceeds via the formation of the lower oxides, the last stage being

$$ZrO + C = ZrC + CO.$$

Thus, the results of the investigation given in [380] are erroneous.

The equilibrium in the system ZrO_2 — C at high temperatures has been investigated in [303, 304].

At temperatures of 1500-1700°C and pressures of 26-133.3 kN/m^2 (150-1000 mm Hg), an oxycarbide is formed with the composition $ZrC_{0.71}O_{0.08}$ and a lattice parameter of 0.4665 nm; the equation of the reaction to which the equilibrium corresponds at these temperatures has the form

$$ZrO_2 + 2.62\,C \rightleftarrows ZrC_{0.7}O_{0.08} + 1.92\,CO.$$

Above 2100°C oxygen is not present in zirconium carbides, but the combined carbon content is lower than the amount calculated for ZrC.

A method of obtaining porous specimens of zirconium carbide by direct sintering during reduction is of interest [939]. By slowly raising the temperature and by adding some fine-grained carbides to the charge, it is possible to obtain specimens with a porosity up to 70% and with pores of uniform size.

Zirconium carbide is produced in large quantities by the carburization of metallic zirconium powder in vacuum.

The diffusion of carbon into zirconium has been studied in [381, 382, 784]. From the determination of the coefficient of diffusion of carbon into β-Zr in the temperature range 900-1260°C, using ^{14}C [784], it has been possible to derive the temperature dependence of the diffusion coefficient, as expressed by the equation

$$D = 0.0048 \exp(-26{,}700/RT)\ \text{cm}^2 \cdot \text{sec.}$$

The method of obtaining zirconium carbide by deposition from the gaseous phase was first used in [314], the carbide being deposited from a gaseous mixture of zirconium tetrachloride, carbon monoxide, and hydrogen.

In [26], zirconium carbide was obtained by passing a mixture of methane and hydrogen through an externally heated retort, on the bottom of which was an ample supply of zirconium chloride heated to evaporation. Above it was a 0.05-mm-diameter filament which acted as carrier. At low temperatures a polycrystalline granular deposit is formed; at high temperatures (depending on the carrier) the deposit is in the form of a single crystal.

Metallographic examination has shown that all the products are homogeneous and have a dense structure.

According to data in [53], in the deposition of zirconium carbide on a tungsten filament, using $C_6H_5CH_3$ as the constituent, the optimum filament temperature is 1730-2430°C. In this process the temperature corresponding to the optimum vapor pressure for $ZrCl_4$ is 300-350°C and for $C_6H_5CH_3$, 15°C.

The data given are in good agreement with the results reported in [50], in which the deposition of zirconium carbide on high-temperature heat-resistant materials was investigated. It was shown there that ZrC is deposited on a tungsten filament from vapors of $ZrCl_4 + H_2 + C_xH_y$ (CH_4, C_6H_6, $C_6H_5CH_3$, C_2H_2, CO, etc.) at atmospheric pressure at 1700-2400°C.

The work described in [229] is interesting; in this the processes of preparing thin zirconium carbide plates were investigated by using compaction followed by activated sintering (with addition of $NiCO_3$ or $Ni_2C_2O_4$) and rolling of the carbide powder into a strip with subsequent sintering.

The System Hafnium—Carbon. According to the hypothetical phase diagram in [371], the system Hf—C contains one carbide corresponding to the composition HfC; its existence was first discovered by Agte and Moers [53, 287].

The phase diagram has been more fully investigated by Benesowsky and Rudy [375], who, on the basis of the results of x-ray

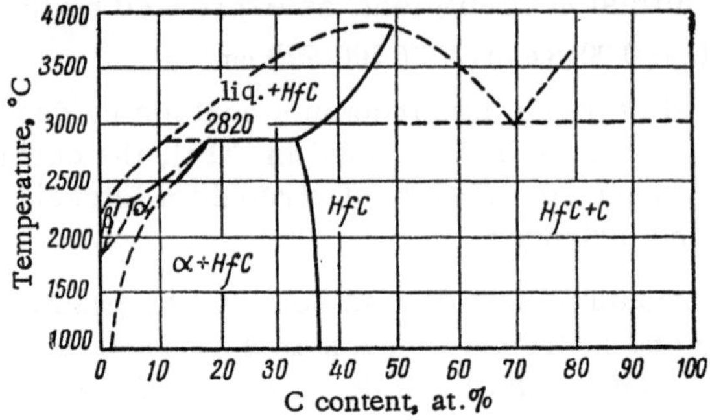

Fig. 45. Phase diagram of the system Hf — C [375].

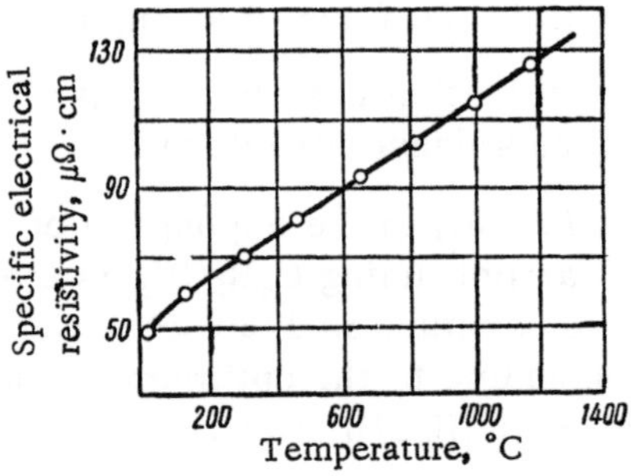

Fig. 46. Temperature dependence of the specific electrical resistivity of HfC [411].

and metallographic analyses and also of melting-point measurements, put forward the diagram shown in Fig. 45. However, measurement of the melting points presented considerable difficulties, and hence they should not be regarded as completely accurate.

According to the data obtained, there is a single phase in the system Hf—C corresponding to the composition HfC_{1-x}, with a homogeneity range extending from 37 to 50 at.% C. The lattice parameter of HfC_{1-x} increases linearly with rise in the carbon content. However, the data on the absolute value of the lattice parameter of the carbide with stoichiometric composition given by various authors [369, 392, 407-409] show discrepancies, the values lying between 0.4614 and 0.4639 nm; this is due to the insufficiently accurate determination of the composition of the specimens and failure to take account of impurities. The HfC — C eutectic is formed

TABLE 15. Lattice Parameters and Coefficient of Thermal Expansion of Hafnium Carbide

Phase	a, nm		Δa, nm	$\alpha \cdot 10^{-6}$ deg^{-1}
	26° C	−190° C		
$HfC_{0.90}$	0.46382±0.00003	0.46347±0.00002	0.00035	3.5±0.4
$HfC_{0.95}$	0.46381±0.00004	0.46338±0.00001	0.00043	4.3±0.4
$HfC_{0.97}$	0.46395±0.00006	0.46354±0.00003	0.00041	4.1±0.7

at about 70 at.% C and a temperature of 3000°C. According to data in [289], the eutectic temperature is 3250 ± 50°C, which is in good agreement with the data in [716].

The results of an x-ray study of hafnium carbides containing various amounts of carbon at −190 and + 26°C are given in Table 15 [392].

The coefficient of linear expansion increases with rise in temperature and is $6.59 \pm 0.04 \cdot 10^{-6}$ deg^{-1} in the range 20–612°C [412].

The temperature dependence of the specific resistivity, based on data in [411] (Fig. 46), indicates the metallic nature of conduction in hafnium carbide.

Measurement at 1130–1860°C of the emission of HfC_{1-x} deposited on tungsten and tantalum has shown [414] that a rise in temperature within this range leads to a work-function value of 3.9 eV ($6.24 \cdot 10^{-19}$ J), which is retained up to 2090°C. On heating the cathode for 3 h at 2130°C, the work function increases to 4.1 eV ($6.56 \cdot 10^{-19}$ J).

The thermodynamic characteristics of hafnium carbide have been studied in [402, 403, 540].

The evaporation rate, composition, and vapor pressure of the carbides have been investigated in specimens of composition $HfC_{0.961}$ [337]. It was shown that the evaporation process takes place in accordance with the equation

$$HfC_{solid} = Hf_{gas} + C_{gas}.$$

The evaporation rate at 2900°C is $(2.5\text{–}6.5) \cdot 10^{-5}$ g/cm^2 · sec.

TABLE 16. Heats of Combustion and Formation of Hafnium Carbides of Various Compositions

Content, %			Formula of compound	Heat of combustion (298°K)		Heat of formation (298°K)	
Hf	C_{comb}	C_{free}		kJ/mole	kcal/mole	kJ/mole	kcal/mole
97.2	2.5	0.1	$HfC_{0.38}$	—1137	—265.8	—151±3.3	—36.0±0.8
96.15	3.8	0.1	$HfC_{0.59}$	—1749	—280.4	—172.2±3.8	—41.1±0.9
95.65	4.3	0.1	$HfC_{0.67}$	—1162	—277.3	—216.6±1.7	—51.7±0.4
94.95	5.0	0.2	$HfC_{0.78}$	—1202	—286.7	—220.8±2.1	—52.7±0.5
94.25	5.75	0.05	$HfC_{0.91}$	—2478	—297.8	—225.0±2.5	—53.7±0.6
93.6	6.05	0.3	$HfC_{0.96}$	—1269	—303.1	—223.3±4.2	—53.3±1.0
94.0	6.25	0.05	$HfC_{0.99}$	—1277	—304.9	—227.0±1.2	—54.2±0.3
93.5	6.3	0.2	HfC	—1282	—305.9	—227.0±2.1	—54.2±0.5

The production of hafnium carbide by the reduction of the dioxide with carbon has been studied in [287, 390, 400-405].

In [402, 403] a dynamic method was used to study the equilibrium of the reduction of the dioxide by carbon at temperatures of 1470-1730°C and pressures of 933.3-13,332 N/m^2 (70-1000 mm Hg).

The dependence of $\log p$ on $1/T$ is expressed by a straight line, thus indicating the practically constant composition of the phases and the monovariance of the equilibrium, which is determined by the equation

$$HfO_2 + 2.9\,C = HfC_{0.95}O_{0.05} + 1.95\,CO.$$

In more recent research [393], the heats of combustion of hafnium carbides of various compositions were determined and their heats of formation calculated (Table 16).

If the process is carried out in an electric resistance furnace with a graphite heater and in a hydrogen atmosphere, the optimum temperature for producing hafnium carbide close to the stoichiometric composition is 2100-2500°C. The product obtained in this way contains 6.26-6.30% C_{total} and up to 0.1% C_{free}.

A study of the conditions for obtaining the carbide by reduction with carbon monoxide in vacuum has shown [404] that oxygen can be completely removed by this process at a temperature of about 2000°C. The use of a charge of stoichiometric composition leads to the formation of products containing less combined carbon than corresponds to the composition HfC; the HfC lattice is retained, but with a reduced parameter, indicating the formation of a lattice deficient in carbon.

The use of a charge containing an excess of carbon black (0.4% in relation to the carbide formed) leads to the formation at 2200°C of a carbide containing 6.26% C_{comb} and 0.22% C_{free}.

A high vacuum leads to the formation of a carbide saturated with carbon. The retention in the reaction space of a certain residual pressure of CO promotes complete saturation of the carbide with carbon.

The conditions for carrying out the reaction $HfO_2 + 3C = HfC + 2CO$ in vacuum have also been investigated in [405] at temperatures of 1000-2000°C under a pressure of 13.33 N/m^2 (0.1

mm Hg). The results show that up to 2000°C a carbide deficient in carbon atoms is formed.

By grinding the products formed at 1900°C and again briquetting and heating at the same temperature for 1 h, it is possible to obtain a carbide whose composition is close to HfC (6.31% C_{comb}).

This two-stage process for producing hafnium carbide at 1900°C is in practice more convenient than holding the charge at 2000°C for 1 h, followed by holding at 2400°C for 1 h and at 3000°C for 5 min, as suggested in [406].

The purest hafnium carbide is obtained by deposition on a 0.05-mm-diameter tungsten wire on heating it in an atmosphere consisting of hafnium tetrachloride vapor, pure nitrogen-free hydrogen, and hydrocarbons of the methane, toluene, and acetylene types [50, 287], the reaction proceeding in accordance with the equation

$$HfCl_4 + CH_4 + H_2 = HfC + 4HCl + H_2.$$

The form of the deposited carbide is determined by the temperature of the wire; the optimum temperature is 2100-2500°C. The high melting point of hafnium makes it possible to obtain the carbide by the direct carburization of a hafnium wire or a filament on which hafnium has previously been deposited, by hydrocarbons at temperatures up to 2500°C [54].

Hafnium carbide powder readily lends itself to hot compaction, forming dense specimens which have a gray metallic lustrous fracture.

In order to examine certain physical properties [405], specimens were prepared from hafnium carbide powder by hot compaction at about 3000°C and a pressure of 17.6 MN/m² (180 kgf/cm²) for 10-15 min. In this way specimens were obtained which had a minimum residual porosity of about 6%.

The equilibrium of the reaction $HfN + C \rightleftharpoons HfC + \frac{1}{2} N_2$ was investigated in [410]. It was shown that the limiting factor in the process is the diffusion of carbon particles into HfN. The equilibrium constant of the reaction in the temperature range 1700-2300°C is given by the equation

$$\log K_p = -\frac{3380}{T} + 2.00.$$

From the results of these investigations it may be concluded that compact specimens of hafnium carbide can be produced in resistance furnaces with graphite heaters at 2500°C in a nitrogen atmosphere. The maximum nitrogen content does not exceed 1% in this method.

Carbides of Metals of Subgroup Va

The System Vanadium—Carbon. According to the first phase diagram, the system vanadium—carbon contains a carbide whose composition varies between $VC_{0.75}$ and VC [371].

Two phases corresponding to the formulas V_4C_3 (15.01% C) and V_5C (4.5% C) were found in the V–C system in [415]. V_5C has a hexagonal close-packed structure. According to [416, 417], the hexagonal structure described belongs to the phase V_2C; V_4C_3 has a fcc lattice [415].

X-ray studies of the alloys having compositions between V_4C_3 and VC have shown [418] that they are all identical and have an NaCl-type fcc lattice. This suggests the existence of a phase VC having a broad homogeneity range, the lower limit of which corresponds to V_4C_3 and the upper limit to VC.

Investigations of the system V—C [419-426] have yielded conflicting data on the limiting composition of the higher carbide and the boundaries of the homogeneity range of the phases V_2C and VC. The most thorough investigation of this system has shown [421, 423] that the following phases are formed by the reaction of vanadium trioxide with carbon:

β-phase of composition $VO_{0.6}C_{0.1}$—$VO_{0.7}C_{0.12}$ having an fcc lattice with parameter a = 0.401 nm;

γ-phase of composition $VC_{0.33}O_{0.09}$—$VC_{0.50}O_{0.13}$ having a hexagonal lattice with parameters a = 0.2857-0.2888 nm, c = 0.4558-0.4567 nm, c/a = 1.592-1.581;

δ-phase whose composition depends markedly on the experimental conditions and whose carbon content reaches 13-14%; it has an fcc lattice with parameter a = 0.4007-0.4123 nm;

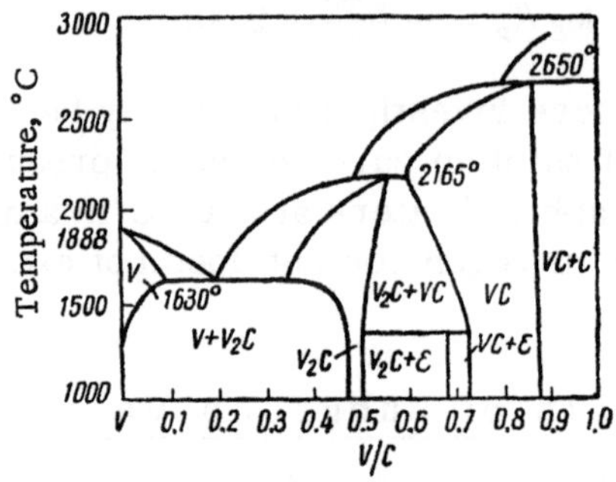

Fig. 47. Phase diagram of the system V−C [425].

ε-phase containing from 14 to 18-19% carbon ($VC_{0.69}-VC_{0.99}$) and having a NaCl-type fcc lattice with parameter a = 0.414-0.415 nm.

The existence of the δ and ε cubic phases has been confirmed in [424]; their composition ranges are $VC_{0.69}-VC_{0.79}$ and $VC_{0.79}-VC_{0.92}$, respectively.

The most thorough investigation of the system V − C has been made in [425], and the phase diagram proposed there is given in Fig. 47. It has been established that the lower limit of the homogeneity range of the phase depends on the temperature. At 1000°C it corresponds to $VC_{0.74}$, at 1800°C to $VC_{0.69}$, and at 2165°C to $VC_{0.60}$. The upper limit corresponds to $VC_{0.88}$; this is in good agreement with [426, 464, 514, 521, 538, 555], in which the authors, in discussing the crystallochemical features of the cubic carbides of titanium, vanadium, and niobium, arrived at the conclusion that the upper boundary of the homogeneity range corresponds to the composition $VC_{0.875}$. More complete filling of the carbon sublattice is unfavorable in terms of energy. With a degree of defectiveness of the sublattice greater than that corresponding to the composition $VC_{0.667}$, the stability of the cubic carbide diminishes, and there is a likelihood of the lower carbon carbide V_2C being formed.

The lattice parameter has been determined by many investigators [11, 22, 173, 331, 384, 422, 435, 436, 864]; for the saturated carbide $VC_{0.88}$ it is 0.4159 nm.

The homogeneity range of the phase V_2C_{1-x} lies between $VC_{0.47}$ and $VC_{0.50}$. A new phase ξ has been found at temperatures below 1334°C in the composition range between $VC_{0.50}$ and $VC_{0.74}$; it is similar to the phase found in the systems Nb — C and Ta — C.

The monocarbide VC forms a eutectic with carbon which melts at 2170 ± 50°C [716]. The monocarbide is a gray metallic powder. According to the majority of investigators, its melting point is 2830°C [25, 85, 435]. The carbide is extremely hard and will scratch corundum.

Measurements of the specific resistivity and Hall constant [855, 864] indicate the metallic nature of conduction in vanadium carbide.

A study of the thermionic-emission properties has shown [414] that the carbide containing 17.5% C_{comb}, 0.15% C_{free}, and 80% V evaporates rapidly at a cathode temperature of 1930°C. At 1330-1430°C the work function of the carbide deposited on tantalum is 3.9 eV, and this value is retained up to 1740°C. Raising the temperature to 1860°C increases the work function to 4 eV.

The evaporation of vanadium carbide has been investigated by Knudsen's method [429]. The composition given for the carbide investigated is $VC_{0.992}$ (18.89% C_{comb}); this raises doubts, since, according to the above data, it is impossible to obtain a carbide of this composition.

The thermodynamic characteristics of vanadium carbide are given in [43, 316, 368, 429, 437, 438, 599].

There are numerous data on the heat of formation of vanadium carbide from its elements, and they show considerable discrepancies. Thus [429] gives a value −95.5 kJ/mole (−22.8 kcal/mole) relating to $VC_{0.992}$; [368] gives −117.3 ± 42 kJ/mole (−28.0 ± 10 kcal/mole) referring simply to VC of various compositions; [43] gives −125.7 kJ/mole (−30.0 kcal/mole) relating to $VC_{0.96}$; [437] gives −171.8 kJ/mole (−43 kcal/mole) relating to $VC_{0.91}O_{0.04}$; [599] gives 101 kJ/mole (−24.1 kcal/mole) relating to $VC_{0.88}$.

Such considerable discrepancies between the data reported arise from insufficient accuracy in determining the composition of the specimens investigated. The most accurate methods were those used in [599].

Vanadium pentoxide or trioxide, ammonium vanadate, and metallic vanadium in powder form may be used as raw materials in producing the carbide. An investigation into the reduction of vanadium oxide by carbon with the object of obtaining metallic vanadium in vacuum is described in [41, 44].

The rate of reduction of V_2O_3 by lamp black at temperatures up to 1000°C is very small, but it increases rapidly between 1200 and 1575°C. The values of the activation energy of the process concerned, based on experimental data in [41], are (204.5–217.5) · 10^3 J/mole (48.8–51.9 · 10^3 cal/mole), which is in good agreement with the activation energy of the reaction $CO_2 + C = 2CO$, namely, 184.4–217.9 kJ/mole (44–52 kcal/mole). This confirms the prevailing view of the nature of the reduction of oxides by carbon, according to which the actual reducing agent is mainly CO [23].

Investigations of the dependence of the CO pressure, temperature, particle size of V_2O_3, compaction pressure, and the presence of niobium, tantalum, chromium, molybdenum, and tungsten powders as impurities on the kinetics of the reduction of V_2O_3 by carbon have shown [821] that the degree of reduction of the oxide increases with rise in temperature from 20% at 1300°C to 68% at 1500°C. The particle size of the V_2O_3 and the compaction pressure do not affect the degree of reduction. A decrease in the CO pressure in the system and additions of refractory-metal powders promote the transformation of the oxycarbide δ'-phase based on VC_{1-x} into the γ'-phase based on V_2C_{1-x}.

In the first work on producing vanadium carbide, V_2O_5 was reduced with carbon [294]. However, it was shown later [25, 431, 432] that it is more efficient to use as the raw material the trioxide V_2O_3, whose melting point is 2000°C as compared with 675°C for the pentoxide V_2O_5.

According to data in [22], a carefully mixed charge consisting of 73–75% V_2O_5 and 27–25% graphite is compressed into briquets, which are heated at 1800°C in a hydrogen atmosphere. The product obtained is heated a second time in vacuum at 1600–1700°C, and as a result a carbide is formed containing 18.5–19% C_{total} and up to 0.5% C_{free}; this value of the C_{free} content is low.

The conditions for producing vanadium carbide in vacuum have been studied most thoroughly in [434].

Investigation of the temperature dependence of the carbide composition has shown that at 1700–1800°C there is no oxygen in the reaction products at any pressure between 13.33 and 1333.2 N/m^2 (0.1 -10 mm Hg). At 1500°C and a pressure of 13.33 N/m^2 (0.1 mm Hg), heating for 2 h is sufficient to completely remove oxygen; the combined carbon content is 17.2-17.3%. Raising the temperature to 1800°C does not lead to any increase in the combined carbon content.

The maximum C_{comb} content does not exceed 17.6-17.8%.

A study has been made in [433] of the conditions for producing vanadium carbide by using as raw material commercial ammonium vanadate containing 56.23% V_2O_5 (theoretical content 56.2%); the reaction proceeds in accordance with the equation

$$2NH_4VO_3 + 5C + 2H_2 = 2VC + 3CO + 2NH_3 + 3H_2O.$$

It was rather difficult to obtain vanadium carbide by deposition on a tungsten filament from vanadium tetrachloride in the presence of hydrogen. Vanadium, which melts at 1712°C, reacts with tungsten at temperatures in this region, thus preventing the formation of the carbide. Hence, the process is first carried out at a lower temperature (1400-1500°C) and only after a film of the carbide has been formed on the filament is the temperature raised to 2000°C [287].

Vanadium carbide has been obtained by a similar method [50] by depositing it on a tungsten filament from a mixture of the tetrachloride, hydrogen, and hydrocarbon.

The System Niobium — Carbon. The first x-ray examination of Nb — C alloy specimens containing 2.5-10.6% C_{comb} was reported in [440]. Lines from two phases with lattice parameters of 0.4425 and 0.4441 nm were observed on the x-ray diffraction diagrams of specimens containing 37 at.% C. These were the phases Nb_2C, containing 33 at.% C, and NbC_{1-x}, containing 41-47 at.% C. It was not possible to obtain the carbide containing 50 at.% C. The existence of a carbide Nb_4C_3 was suggested in [441].

The system Nb — C has been studied in detail in [442-449].

The results of the investigations, as given in [443], are reproduced in Table 17.

Subsequent researches on Nb — C alloys [446-448] enabled the phase diagram of this system to be constructed. The existence of three single-phase regions in the system was confirmed:

TABLE 17. Characteristics of Specimens in the System Nb — C

Content, wt.%				Formula of compound	Phases observed	Lattice parameters, nm			
Nb	C_{total}	C_{free}	$Nb+C_{total}$			α-phase	β-phase		γ-phase
						a	*a*	*c*	*a*
99.12	0,79	—	99.91	$NbC_{0.06}$	α, β	0.33011	—	—	—
98.15	1,82	—	99.97	$NbC_{0.14}$	α, β	0.33012	—	—	—
96.95	3.00	—	99.95	$NbC_{0.24}$	β, α	—	0.3119	0.4957	—
95.99	4.26	—	99.95	$NbC_{0.34}$	β, α	—	0.3120	0.4957	—
94.83	5.19	—	100.02	$NbC_{0.42}$	β	—	0.3121	0.4960	—
94.04	5.94	—	99.98	$NbC_{0.49}$	β	—	0.3123	0.4966	—
93.48	6,45	—	99.93	$NbC_{0.53}$	β, δ, (ξ)	—	0.3128	0.4974	—
92.78	7.26	—	100.04	$NbC_{0.60}$	β, δ, (ξ)	—	0.3128	0.4974	0.44317
92,34	7.61	—	99.95	$NbC_{0.64}$	δ, β, (ξ)	—	—	—	0.44300
92.03	7.93	—	99.96	$NbC_{0.66}$	δ, β, (ξ)	—	—	—	0.44315

91.72	8.28	—	100.00	$NbC_{0.70}$	δ	—	—	—	0.44309
91.59	8.34	—	99.93	$NbC_{0.71}$	δ	—	—	—	0.44335
91.26	8.61	—	99.87	$NbC_{0.73}$	δ	—	—	—	0.44395
90.90	9.12	—	100.02	$NbC_{0.78}$	δ	—	—	—	0.44519
90.01	9.95	—	99.96	$NbC_{0.85}$	δ	—	—	—	0.44618
89.80	10.15	—	99.95	$NbC_{0.87}$	δ	—	—	—	0.44649
89.70	10.24	—	99.94	$NbC_{0.88}$	δ	—	—	—	0 44660
89.50	10.51	—	100.01	$NbC_{0.91}$	δ	—	—	—	0.44690
89.29	10.66	—	99.95	$NbC_{0.92}$	δ	—	—	—	0.44691
89.21	10.72	0.02	99.93	$NbC_{0.93}$	δ, graphite	—	—	—	0.44682
88.94	11.01	0.23	99.95	$NbC_{0.95}$	δ, graphite	—	—	—	0.44692
88.57	11.38	0.53	99.95	$NbC_{0.99}$	δ, graphite	—	—	—	0.44690

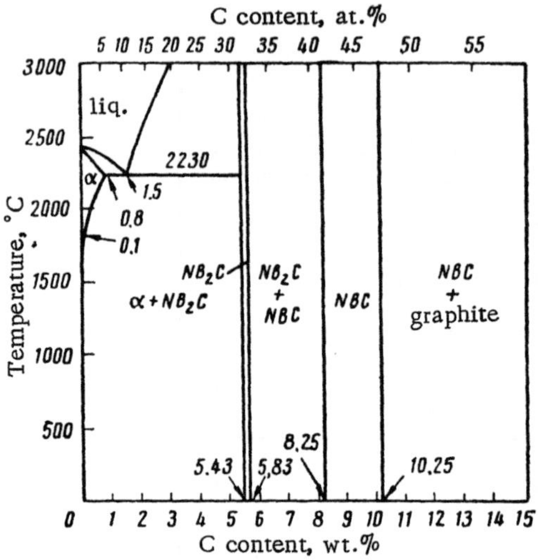

Fig. 48. Phase diagram of the system Nb — C [449].

α -phase — a solid solution of carbon in niobium. According to the proposed diagram, the solubility of carbon in niobium decreases with fall in temperature;

β -phase — Nb_2C_{1-x}, melting peritectically at 3090 ± 500°C, with a homogeneity range from $NbC_{0.495}$ — $NbC_{0.500}$ below 2000°C; the α - and β -phases form a eutectic melting at 2335 ± 20°C [448] or 2328 ± 17°C [443];

γ -phase — the carbide NbC_{1-x} with a peritectic temperature of 3500 ± 75°C for specimens of composition $NbC_{0.86}$. Its homogeneity range extends from $NbC_{0.71}$ to $NbC_{0.90}$ below 2000°C. There is a $\beta + \gamma$ two-phase region between the β - and γ-phases.

No evidence was found of the ξ -phase reported in [443] in the composition range $NbC_{0.50}$ — $NbC_{0.70}$. The NbC — C phase boundary lies in the vicinity of $NbC_{0.91}$ at 2000°C.

The system Nb — C was investigated in [449] by x-ray and metallographic analyses of alloys produced by arc melting and by sintering niobium powder with lamp black. The specimens contained 0.011-14.12% C.

Like the previous investigations, the results, which are presented in the form of a phase diagram in Fig. 48, reveal the existence of two carbides in the Nb–C system: a hexagonal carbide Nb_2C_{1-x} with a homogeneity range of 5.48-5.83 wt.% C ($NbC_{0.44}-NbC_{0.48}$), and a fcc carbide NbC_{1-x} with a homogeneity range of 8.25-10.25 wt.% C ($NbC_{0.70}-NbC_{0.90}$). However, the phase boundaries of the homogeneity ranges of the two carbides, as reported by various authors, differ considerably.

The most probable concentration limits for the stable carbide phases are $NbC_{0.48}-NbC_{0.50}$ for the hexagonal carbide and $NbC_{0.70}-NbC_{0.99}$ for the cubic one [444].

The maximum solubility of carbon in niobium is 0.8% at the eutectic temperature (2230°C), decreasing rapidly with temperature to 0.1 wt.% at 1800°C.

An investigation of the reaction of the monocarbide with carbon has shown [289] that the system is of the eutectic type, the melting point of the eutectic lying at 3150 ± 50°C; this is in good agreement with the data in [716]. An x-ray analysis of solidified drops of the eutectic revealed the presence of two phases: NbC_{1-x} and graphite.

From an investigation of the diffusion of carbon into the carbide it has been possible to determine the diffusion coefficient $D = 8 \times 10^{-7}$ cm^2/sec at 2670°C.

Gel'd and Lyubimov [454] have studied the self-diffusion of niobium, the diffusion of carbon into metallic niobium, and its self-diffusion in the carbide. The self-diffusion of niobium was studied on a specimen of composition $NbC_{0.98}$, and the temperature dependence of the mobility of niobium is represented by the equation

$$D = 5.5 \cdot 10^{-6} \exp\left(-\frac{55{,}000}{RT}\right).$$

The activation energy of self-diffusion of niobium in $NbC_{0.98}$ is 35% less than that in metallic niobium.

The temperature dependence of the diffusion coefficient of carbon into niobium is represented by the equation

$$D = 1.09 \cdot 10^{-5} \exp\left(-\frac{32{,}000}{RT}\right).$$

TABLE 18. Lattice Parameter and Coefficient of Linear Expansion of $NbC_{0.884}$

Property	Temperature, °C									
	20	498	590	700	720	848	1000	1500	2000	2500
a, nm	0.44691 ± 0.00003	0.44833 ± 0.00003	—	0.44895 ± 0.00004	—	0.44943 ± 0.00004	—	—	—	—
$\alpha_{298}^{T} \cdot 10^{-6}$ deg^{-1}	—	6.65 ± 0.28	6.70	—	6.76 ± 0.23	6.84 ± 0.21	7.11	8.10	8.27	8.75

The diffusion of carbon in the carbide has been studied on specimens of $NbC_{0.5}$, $NbC_{0.75}$, and $NbC_{0.88}$; the temperature dependence of the self-diffusion coefficients is described by the equations

$$D^{c}_{NbC_{0.5}} = 5.7 \cdot 10^{-6} \exp\left(-\frac{29,800}{T}\right),$$

$$D^{c}_{NbC_{0.75}} = 8.8 \cdot 10^{-6} \exp\left(-\frac{35,300}{T}\right),$$

$$D^{c}_{NbC_{0.98}} = 1.06 \cdot 10^{-6} \exp\left(-\frac{32,300}{T}\right).$$

The activation energies of the self-diffusion of carbon in niobium carbides of various compositions depend very little on the composition.

The higher carbide NbC_{1-x} is a cinnamon-gray metallic powder. According to data in [456], the distance Nb — C is 0.22355 nm. There are discrepancies between the values of the lattice parameters reported by various authors [439–444].

From a determination of the lattice parameter of specimens of NbC_{1-x} with a molar ratio Nb : C = 0.700 : 0.996, it has been possible to derive [445] the dependence of the lattice parameter on this ratio; it is expressed by the equation

$$a = 0.409847 - 0.071820\,(\mathrm{C/Nb}) - 0.034570\,(\mathrm{C/Nb})^2.$$

Table 18 gives the values of the lattice parameters of the carbide $NbC_{0.884}$ at high temperatures [335] and the calculated values of the coefficient of thermal expansion.

A study of the specific electrical resistivity and the Hall constant [19, 855, 948] has confirmed the metallic nature of the conduction in NbC_{1-x}.

An investigation of the evaporation of specimens of composition $NbC_{0.976}$ in the temperature range 2500–2854°C has shown [462] that dissociation takes place in accordance with the equation $NbC_{solid} = Nb_{gas} + C_{gas}$.

The evaporation of NbC has been studied in [463] by Langmuir's method over the temperature range 1990–2670°C. It was shown that on heating the carbide $NbC_{0.97}$ in vacuum at 2670°C the ratio C/Nb falls until it reaches a value corresponding to the composition $NbC_{0.75}$, after which it becomes constant.

The vapor pressure of carbon over NbC_{1-x} close to the stoichiometric composition is represented by the equation

$$\log p_C = 5.296 - \frac{32\,760}{T}.$$

The calculated heat of evaporation is 628 kJ/mole (150 kcal/mole).

From a study of the thermodynamic properties of niobium carbides of various compositions [457, 458], equations have been derived for the temperature dependence of the specific heat and the change in enthalpy for the carbides $NbC_{0.749}$, $NbC_{0.867}$, and $NbC_{1.00}$.

The temperature dependence obtained for the change in enthalpy with composition is given in Fig. 49.

The heat of formation of 1 g-atom increases with fall in carbon content, the relationship being well expressed by the equation

$$\Delta H_{NbC_x} = 4.19\,(48.1\,x - 15{,}9\,x^2)\ \text{kJ/mole}$$

$$[\Delta H_{NbC_x} = (48.1\,x - 15.9\,x^2)\ \text{kcal/mole}].$$

The heats of formation given in various papers [339, 457–460, 462, 599] show discrepancies. In a number of cases the authors do not even give the composition of the specimens examined; sometimes the composition has not been checked with sufficient accuracy and the impurity content has not been determined. The thermodynamic properties of niobium carbides of various

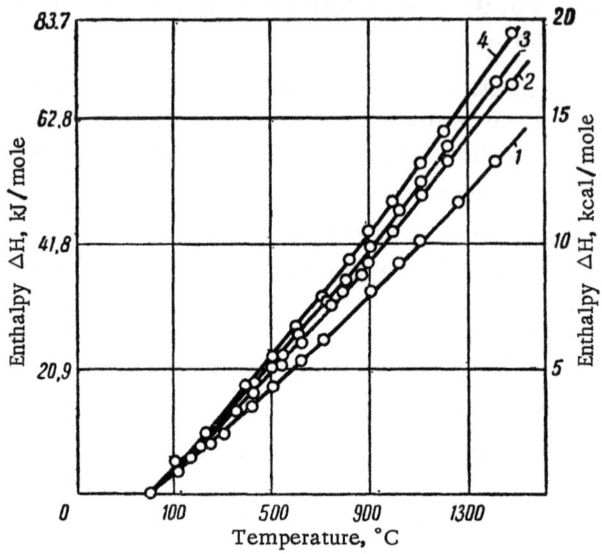

Fig. 49. Temperature dependence of the enthalpy variation of niobium carbides of compositions: 1) $NbC_{0.50}$; 2) $NbC_{0.749}$; 3) $NbC_{0.867}$; 4) $NbC_{1.00}$.

TABLE 19. Values of the Standard Heats of Combustion and Heats of Formation of Niobium Carbides

Carbide	Lattice param. a, nm	Heat of combustion (298°K)		Standard heat of formation ΔH^0_{form} (298°K)	
		kJ/g formula	kcal/g formula	kJ/g formula	kcal/g formula
$NbC_{0.739}$	0.4442± ±0.0001	−1121.2± ±1.7	−267.6± ±0.4	−120.2± ±2.1	−28.7± ±0.5
$NbC_{0.783}$	0.4454± ±0.0001	−1137.6± ±2.5	−271.5± ±0.6	−121.1± ±2.9	−28.9± ±0.7
$NbC_{0.838}$	0.4458± ±0.0001	−1154.8± ±1.7	−275.6± ±0.4	−125.7± ±2.1	−30.0± ±0.5
$NbC_{0.913}$	0.4467± ±0.0001	−1179.9± ±2.1	−281.6± ±0.5	−129.9± ±2.5	−31.0± ±0.6

compositions were thoroughly investigated by precision methods in [460, 461].

To determine the thermodynamic properties more accurately, the heats of combustion of carbides of various compositions were measured in [460]. By the use of data obtained from the heats of combustion and the data obtained in [461] for the heat of formation of Nb_2O_5, it has been possible to calculate the heats of formation of various niobium carbides (Table 19).

The most widely used method of producing niobium carbide is by reduction of the oxide with carbon black in a hydrogen atmosphere or in vacuum.

A study of the conditions and mechanism of obtaining the carbide by reducing Nb_2O_5 with carbon in vacuum has shown [43] that the optimum temperature for producing niobium carbide close to the stoichiometric composition is 1800°C; in this method the time of heating required for completion of the reaction is 4 h. With a shorter time of heating, a considerable part of the carbon and niobium remains in the free state.

The equilibrium in the system Nb — C — O has been investigated in [450], where it was shown that the reduction of NbO to the higher carbide NbC is most probable from the thermodynamic point of view; the conditions for the reduction to Nb_2C are somewhat less favorable, and they are least favorable for the formation of metallic niobium.

An investigation into the effect of the structure of the Nb_2O_5, the pressure, kind of reducing agent, and additions of alkali-metal salts on the carbothermal reduction of the pentoxide has shown [451] that the reaction depends to a considerable extent on the method of preparing the Nb_2O_5. Oxides obtained by heating niobium hydroxide at 800-900°C, which contain mainly the low-temperature T-form of Nb_2O_5, are more reactive than those heated at higher temperatures and containing the medium- and high-temperature M- and H-modifications. This is evidently due to recrystallization of the Nb_2O_5 on heating and to a reduction in the reactive surface.

A study of the effect of pressure on the reduction process (Fig. 50) leads to the conclusion that with rise in pressure from high vacuum to 399.9-533.2 N/m^2 (3-4 mm Hg) the reaction rate increases, but that with further rise in pressure the reaction is retarded. This indicates the different effects of the gas phase on the characteristics of the reduction stage and on the regeneration of CO, which leads to an extreme dependence law.

The type of reducing agent, its grain size (Fig. 51), and additions of alkali-metal salts have a great effect on the reduction process [451, 452]. When acetylene soot, graphite, coal-tar pitch, and its coking products were used as reducing agents, the best results were obtained from the pitch. This is due to the fact that on heating the charge pyrolysis of the hydrocarbons contained in the pitch took place with the formation of soot-like carbon having a large specific surface and a high reaction capacity.

The addition of 2 wt.% K_2CO_3 to the charge raised the degree of reduction of Nb_2O_5 at 1150°C by more than twice.

The second method of preparing niobium carbide is by synthesis from the elements at 1700-1800°C. The raw material consists of metallic niobium powder or niobium hydride, which decomposes at 800-900°C with the evolution of hydrogen.

An attempt to obtain niobium carbide by deposition on a tungsten filament from a mixture of niobium chloride, hydrogen, and hydrocarbons (toluene vapor, methane, acetylene) did not give positive results [50, 287]. At 900-1000°C metallic niobium is rapidly deposited on the filament. However, on further heating in a hydrogen atmosphere containing hydrocarbons at 1400-1700°C, the deposition products could be completely converted into the carbide.

The production of parts of various shapes from niobium carbide is very difficult owing to the high sintering temperature of the compacts (3000°C). The sintering temperature can be reduced by making additions which activate the sintering process. In [455] conditions have been determined for the activated sintering of niobium carbide containing additions of iron and cobalt chloride; by this means it is possible to obtain parts with a porosity of 15-18% by sintering for 1.5 h at 1500°C. Sintering specimens under the same conditions but without the activating additions results in parts with a porosity of 30-40%.

The System Tantalum—Carbon. The system Ta — C was first studied in [315], in which the existence of two carbides was established: Ta_2C_{1-x} and TaC_{1-x}. Two modifications

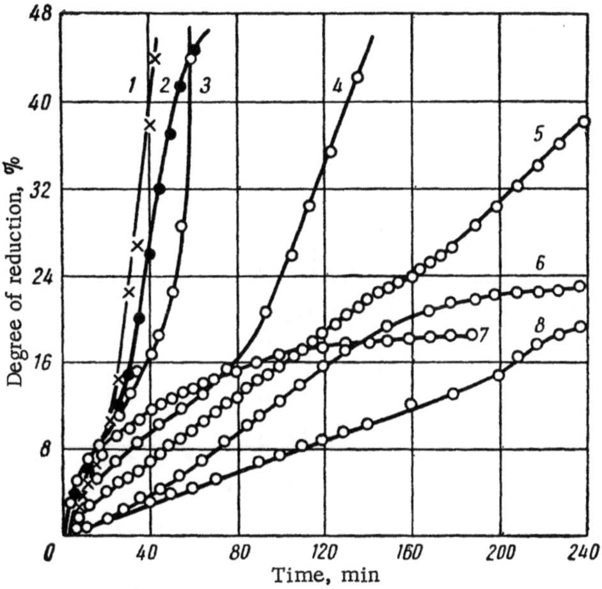

Fig. 50. Kinetics of reduction of Nb_2O_5 at various pressures of the gaseous phase (N/m^2). 1) 13.33; 2) 533.28; 3) 0.13; 4) $13.33 \cdot 10^2$; 5) $26.63 \cdot 10^2$; 6) $79.99 \cdot 10^2$; 7) accumulation of reaction products up to $p_2 = 79.99 \cdot 10^2$; 8) $13.33 \cdot 10^3$.

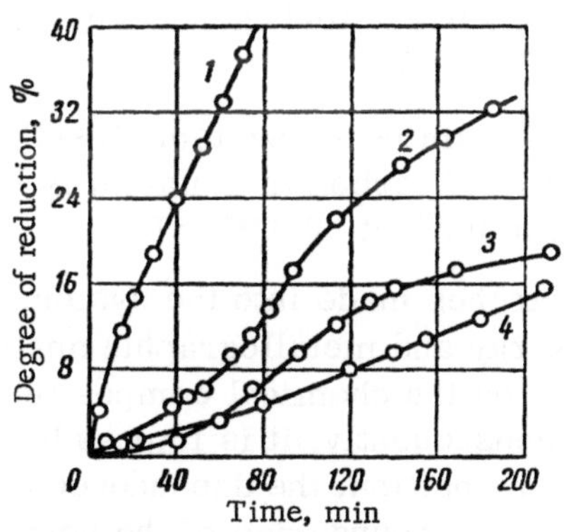

Fig. 51. Dependence of the degree of reduction of Nb_2O_5 on the time reducing agents of various grain sizes used: 1) carbon black; 2-4) graphite with particle sizes (mm); 2) 0.10; 3) 0.25; 4) 0.5.

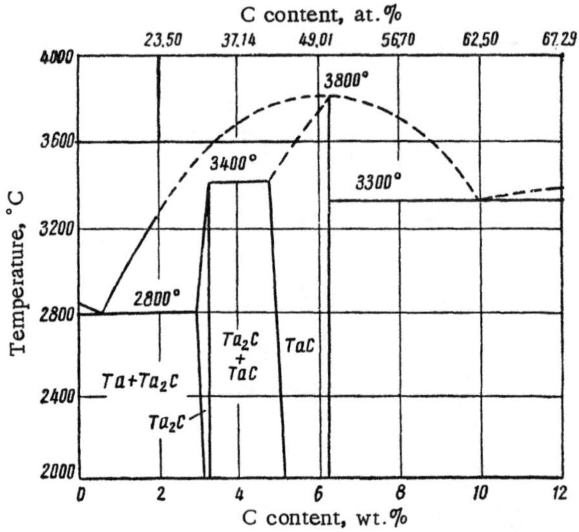

Fig. 52. Phase diagram of the system Ta–C [465].

of the carbide Ta_2C are known: the α-modification, which crystallizes in a hexagonal close-packed lattice with parameters a = 0.3085 ± 0.0001 nm and c = 0.442 ± 0.0007 nm, c/a = 1.59, and the β-modification, whose crystal structure has not been determined though its x-ray diffraction diagram is known to differ very little from that of the α-modification.

The system has been most thoroughly investigated in [283, 289, 315, 384, 447, 465–469].

The phase diagram of the system Ta – C is shown in Fig. 52 [465]. The solubility of carbon in tantalum is less than 0.1% at 1800°C and less than 0.2% at 2000°C [873].

All the researches made into the system Ta – C have consisted mainly of x-ray and metallographic analyses; the authors do not provide data on the chemical composition of the specimens investigated and, consequently, it is impossible to estimate accurately the phase composition, the dependence of lattice parameters on composition, and the boundaries of the homogeneity ranges.

The system has been investigated by means of the most refined methods of x-ray, chemical, and thermodynamic analyses in [467–469].

The following phases have been found:

α -phase is cubic, in which carbon is practically insoluble; it is represented by the composition $Ta_1C_0 - Ta_1C_{0.03}$, with lattice parameters from 0.33035 to 0.33066 ± 0.00002 nm;

β-phase is hexagonal with a homogeneity range from $TaC_{0.38\pm0.02}$ to $TaC_{0.50\pm0.02}$ and lattice parameters from $a = 0.31012$ and $c = 0.49372$ nm for $TaC_{0.38}$ to $a = 0.31042$ and $c = 0.49410$ nm for $TaC_{0.50}$;

γ -phase is fcc with a homogeneity range from $TaC_{0.58\pm0.02}$ to $TaC_{0.91\pm0.02}$ with a lattice parameter from 0.44206 to 0.44564 nm.

The carbide of composition TaC was not obtained by these authors.

The existence of the α -, β -, and γ -phases was confirmed in [466]. The homogeneity ranges and lattice parameters differ somewhat from those in [467–469].

The α -phase extends from Ta to $TaC_{0.03}$ with lattice parameters from 0.33028 to 0.33050 nm. The β -phase extends from $TaC_{0.41}$ to $TaC_{0.5}$ with parameters $a = 0.3101$, $c = 0.4933$ nm to $a = 0.3106$, $c = 0.4945$ nm. The γ -phase extends from $TaC_{0.74}$ to $TaC_{1.00}$.

In addition, a fourth phase (ξ) of composition $TaC_{0.64}$ has been found; it has no homogeneity range and is stable at temperatures below 1800°C.

According to data in [470], the maximum content of combined carbon in tantalum carbide prepared in the temperature range 1500–2500°C corresponds to the formula $TaC_{0.96}$.

The discrepancies in the data for the concentration limits of the homogeneity ranges of the β - and γ -phases have led to still another investigation of the system Ta — C [136]. As a result it has been established that the hexagonal β -phase is a solid solution based on Ta_2C; its homogeneity range extends from $TaC_{0.43}$ to $TaC_{0.50}$, in good agreement with the data in [466]. The homogeneity range of the γ -phase extends from $TaC_{0.72}$ to $TaC_{0.96}$, also in good agreement with the data in [466]. The maximum combined carbon content in the γ -phase corresponds to the composition $TaC_{0.962}$.

The lattice parameter of tantalum carbide has been determined by numerous investigators [294, 315, 329, 384, 385, 465, 469, 472, 478, 479].

Data on the lattice parameters and x-ray density of tantalum carbide in the homogeneity range are given in Table 20.

The lattice parameter increases linearly with rise in carbon content. For TaC of limiting composition the data were obtained by extrapolation, since it was impossible to reach this composition. It has been concluded that the value $a = 0.4455$ nm which is most frequently found in the literature for the carbide TaC in fact corresponds to the compositions $TaC_{0.80}$ and $TaC_{0.85}$, which are the most easily produced. The density of the monocarbide falls with increase in carbon content, the maximum occurring at the composition $TaC_{0.85}$. Values of the lattice parameters of the carbide of composition $TaC_{0.984}$ and the coefficient of linear expansion at temperatures from 23 to 990°C are given in [335] (see Table 21). These data also probably relate to a composition of lower carbon content than $TaC_{0.984}$.

The values of the coefficient of linear expansion given in Table 21 are confirmed by the data in [301, 409].

The temperature dependence of the specific electrical resistivity of tantalum carbide obtained by the reaction of tantalum powder with acetylene in a hydrogen and argon atmosphere at 2000°C [133] is represented by the equation

$$\rho = [102 + 0.0178\,T\,(°\mathrm{K})]\;\mu\Omega\cdot\mathrm{cm},$$

which is valid over the temperature range 1400–3500°K.

The current-carrier concentration calculated from measurements of the specific electrical resistivity and Hall constant [855] is 2.2 electrons/formula.

An investigation of the electrophysical properties of Ta — C alloys containing 4.90–6.16 wt.% of combined carbon, in the temperature range −196 to 155°C, has shown [926] that the specific electrical resistivity increases from 11.6 $\mu\Omega\cdot$cm at 16% C to 147 $\mu\Omega\cdot$cm at 5.18% C.

The thermodynamic properties of tantalum carbides of various compositions were determined in [468, 469, 940] and the

TABLE 20. Lattice Parameters and Density of Tantalum Carbides

Formula of compound	Lattice param., nm	Volume of unit cell $v = a^3$, nm^3	X-ray density, g/cm^3
$TaC_{0.58}$	0.4420	0.08635	11.44_5
$TaC_{0.80}$	0.4444	0.08776	14.41_5
$TaC_{0.90}$	0.4455	0.08842	14.39_8

TABLE 21. Values of the Lattice Parameters and Coefficients of Linear Expansion of $TaC_{0.984}$ at Various Temperatures

Property	Temperature, °C					
	23	498	590	705	848	994
a, nm	0.44540± ±0.00003	0.44673± ±0.00003	—	0.44738± ±0.00004	—	0.44827± ±0.00003
$\alpha_{298}^{T} \cdot 10^{-6} \cdot deg^{-1}$	—	6.29± ±0.29	6.41	6.52± ±0.23	6.61	6.64± ±0.14

microhardness and electrical resistivity in [480] (Tables 22 and 23).

The data available in the literature on the heats of formation of tantalum carbides are very numerous [129, 337, 339, 447, 468]. However, for the most part, they refer to products whose composition is not given at all or has been determined with inadequate accuracy. The most accurate method of determining the heat of formation of tantalum carbides in relation to their composition was used in [129]. The specimens were subjected to careful x-ray and chemical analyses, the content of the following impurities being determined in addition to the principal components: O, N, H, Si, Ti, Fe, Mg, Mn, Ca, Cu, Sn, Zr, Ni, Al, Mo, W, Pb, Bi, Sb, and Cd. The heats of combustion of the carbides TaC_x, where the value of x varied between 0.455 and 0.988, have been determined by a calorimetric method, and on the basis of the values obtained the heats of formation have been calculated, taking into account the heats of formation of Ta_2O_5 and CO_2 (Table 24); the values are in good agreement with the data in [447].

The evaporation of tantalum carbide has been studied on a specimen of composition $TaC_{0.996}$ in the temperature range 2700-

TABLE 22. Properties of Tantalum Monocarbide in the Homogeneity Range

Formula of compound	ΔH_{298}		ΔS_{298}		ΔF_{298}		Microhardness		Specific electrical resistivity, $\mu\Omega \cdot cm$
	kJ/mole	kcal/mole	J/mole · deg	cal/mole · deg	kJ/mole	kcal/mole	MN/m^2	kgf/mm^2	
$TaC_{0.60}$	—78.4	—18.7	—2.01	—0.48	—77.9	—18.6	11110	1111	134.0
$TaC_{0.65}$	—	—	—	—	—	—	—	—	114.0
$TaC_{0.70}$	—90.1	—21.5	—1.80	—0.43	—89.7	—21.4	—	—	126.0
$TaC_{0.75}$	—	—	—	—	—	—	13600	1360	110.8
$TaC_{0.80}$	—106.8	—25.5	—2.10	—0.50	—106.4	—25.4	—	—	—
$TaC_{0.85}$	—	—	—	—	—	—	14460	1446	41.2
$TaC_{0.90}$	—132.8	—31.7	—3.35	—0.80	—132.0	—31.5	—	—	—
$TaC_{0.95}$	—	—	—	—	—	—	14700	1470	34.0

TABLE 23. Some Properties of Ta_2C_{1-x} in the Homogeneity Range

Formula of compound	Lattice parameters, nm		Density, g/cm^3	ΔH_{298}		ΔS_{298}		ΔF_{298}		Microhardness		Specific electrical resistivity, $\mu\Omega \cdot cm$
	a	*c*		kJ/mole	kcal/mole	J/deg · mole	cal/deg·mole	kJ/mole	kcal/mole	MN/m^2	kgf/mm^2	
$TaC_{0.38}$	0.31012	0.49372	14.97	—	—	—	—	—	—	—	—	—
$TaC_{0.40}$	—	—	—	—64.9	—15.5	—3.18	—0.76	—64.1	—15.3	7940	810	138.5
$TaC_{0.45}$	—	—	—	—	—	—	—	—	—	8720	890	136.0
$TaC_{0.50}$	0.31042	0.49410	15.050	—71.2	—17.0	—2.89	—0.69	—70.4	—16.8	9280	947	80.25

TABLE 24. Heats of Combustion and Formation of Tantalum Carbides

Formula of compound	[447] Heat of combustion, kJ/kg	[447] Heat of combustion, cal/g	[447] Heat of formation $-\Delta H^0_{298}$, kJ/mole	[447] Heat of formation $-\Delta H^0_{298}$, kcal/mole
Ta	—5627	—1343	—	—
$TaC_{0.485}$	—5975	—1426	95.9±7.1	22.9±1.7
$TaC_{0.724}$	—6277	—1498	113.5±5.4	27.1±1.3
$TaC_{0.749}$	—6285	—1500	120.2±2.1	28.7±0.5
$TaC_{0.802}$	—6310	—1506	132.0±5.0	31.5±1.2
$TaC_{0.821}$	—6394	—1526	121.5±4.6	29.0±1.1
$TaC_{0.838}$	—6373	—1521	131.0±2.1	31.3±0.5
$TaC_{0.904}$	—6465	—1543	134.9±2.5	32.2±0.6
$TaC_{0.936}$	—6469	—1544	144.1±5.4	34.4±1.3
$TaC_{0.958}$	—6566	—1567	132.0±7.1	31.5±1.7
$TaC_{0.982}$	—6557	—1565	146.6±2.5	33.8±0.6
$TaC_{0.988}$	—6566	—1567	145.0±5.0	34.6±1.2

Formula of compound	[129] Heat of combustion, kJ/kg	[129] Heat of combustion, cal/g	[129] Heat of formation $-\Delta H^0_{298}$, kJ/mole	[129] Heat of formation $-\Delta H^0_{298}$, kcal/mole
—	—	—	—	—
$TaC_{0.734}$	—1200± ±6.3	—286.4± ±1.5	114.4±6.3	27.3±1.5
—	—	—	—	—
$TaC_{0.792}$	—1208± ±3.8	—288.4± ±0.9	125.7±4.2	30.0±1.0
$TaC_{0.826}$	—1227± ±9.2	—292.8± ±2.2	123.6±9.2	29.5±2.2
—	—	—	—	—
—	—	—	—	—
—	—	—	—	—
$TaC_{0.961}$	—1268± ±3.8	—302.7± ±0.9	135.3±4.2	32.3±1.0
$TaC_{0.982}$	—1270± ±3.8	—303.3± ±0.9	141.2±4.2	33.7±1.0

3100°C [337]. It was shown that on heating in a vacuum of 1.33 · 10^{-3} N/m^2 (~10^{-5} mm Hg) in the above temperature range, the ratio C/Ta decreases to a value corresponding to the composition $TaC_{0.77}$, after which it remains constant.

The polishing capacity of tantalum carbide phases in relation to their composition has been determined in [469], and compared with the values for corundum and carborundum. The results indicate an increase in the polishing capacity with rise in carbon content in the TaC_{1-x} phase and show that its value is higher than that of corundum.

An investigation of the polishing capacity of the Ta_2C_{1-x} phase showed it to be lower than that of the monocarbide; it increases with the carbon content within the homogeneity range.

Tantalum carbide has been obtained by heating a mixture of the pentoxide with carbon black in a hydrogen atmosphere at 1200-1600°C [25, 187, 471].

The optimum conditions for obtaining the carbide by this method [472] are to raise the temperature to 1600°C in the course of 1 h and to hold the mixture at this temperature for 2.5 h in a hydrogen atmosphere.

The results of an investigation into the conditions for producing tantalum carbide in vacuum indicate that at 1400-1500°C and with a holding time of 1 h, a product is formed which is similar in composition and lattice parameter to stoichiometric TaC.

The method of preparing tantalum carbide by synthesis from its elements has been used by many investigators [287, 465, 466, 472]. In [472], experiments were carried out in an electric resistance furnace with graphite heaters in a hydrogen atmosphere at 1400-1600°C. The charge consisted of tantalum obtained by the reduction of potassium fluorotantalate with metallic sodium and a calculated amount of carbon black.

In this case, the optimum conditions were to raise the temperature to 1600°C in 1 h and then to hold the charge at this temperature for 2.5 h.

The conditions for producing TaC_{1-x} from pieces of scrap tantalum sheet in a carbon container were also investigated. The container was placed in a carbon boat and packed with pieces of carbon. The optimum conditions were to raise the temperature to 1800°C in 40 min and to hold the charge at this temperature for 3 h.

The production of tantalum carbide by deposition from the gaseous phase [287] is attended by difficulties, owing to the fact that at 900–1000°C metallic tantalum is deposited from a mixture consisting of tantalum chloride vapor, hydrogen, and hydrocarbon; this prevents the carbide from being formed.

In [315], a carbon filament was heated in a mixture of tantalum pentachloride and hydrogen. The pentachloride is reduced at the filament surface to metallic tantalum, which reacts with the hydrocarbons formed to give the carbide. At temperatures below 2000°C, tantalum predominates in the reaction products, while at 2000–2500°C tantalum carbide predominates.

Pure tantalum carbide has been deposited on a tungsten filament from the pentachloride [52, 54]. The mixture of tantalum and carbide formed during deposition was heated in a hydrocarbon atmosphere at 2330–2930°C, and as a result the single-phase carbide was obtained [477].

The conditions under which the tantalum carbide deposited on the filament is stable were investigated in [473]. Since hydrocarbons and hydrogen were present in the atmosphere, lamps in which tantalum monocarbide acted as the light source lasted for more than 50 h at 3200°C.

Owing to the high concentration of hydrocarbons, carbon is deposited in the free form at high temperatures; the equilibrium concentration of the hydrocarbons then decreases, thus causing decarburization of the tantalum carbide.

The addition of hydrogen to the hydrocarbon atmosphere somewhat reduces the rate at which free carbon is deposited. The introduction of bromine or chlorine vapor reduces the amount of hydrogen required to prevent decarburization from a pressure of 26.66 to 6.66 kN/m^2 (from 200 to 50 mm Hg).

Methods are known by which tantalum carbide can be produced from tantalum-containing slags. These slags are subjected to a reducing melt with a large excess of carbon in the presence

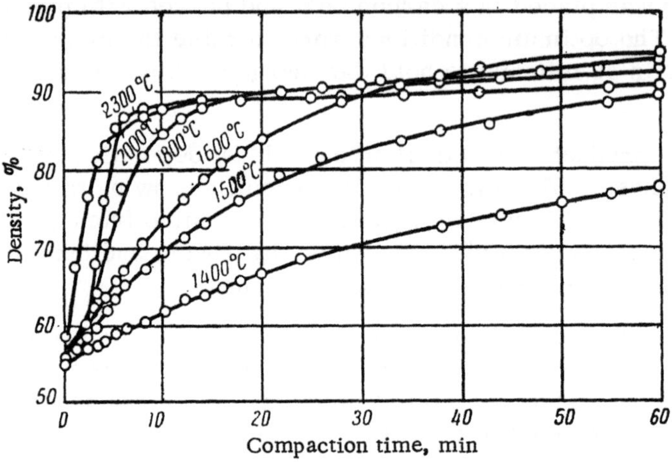

Fig. 53. Dependence of the density of tantalum carbide specimens on the time and temperature of hot compaction.

of iron. The carbide formed is separated chemically. Tantalum carbide mixed with niobium carbide is obtained by the carburization of ores [22] or ferrotantalum [474].

Tantalum carbide has also been obtained from metallic tantalum and carbon in a bath of molten aluminum superheated to 2000°C in a graphite crucible [55]. Metals of the iron group may also be used for the liquid bath. The ingots obtained were treated with acid to dissolve the carbides of aluminum, iron, etc., that had been formed.

An investigation into the effects of temperature, compaction time, and metallic additions on the density of specimens produced by hot compaction [475, 476] has shown that densification of the carbide already takes place at 1400°C, the effect of temperature on the density of the specimens being considerably greater than that of compaction time (Fig. 53). A rapid increase in density up to 95% of the theoretical takes place in the temperature range 1600-2300°C. The process is considerably activated by adding to the charge up to 1% of nickel, cobalt, manganese, or iron.

Specimens produced by compacting carbide powder containing 1% nickel at 1600°C and maintaining them at this temperature for 40-60 min under a pressure of 30 MN/m^2 (300 kgf/cm^2) have a residual porosity of 2-3%; on compacting a mixture containing 1% manganese at 1800°C and holding the specimens for 30-40 min under the same pressure, the residual porosity is 1-2%.

Carbides of Metals of Subgroup VIa

The System Chromium—Carbon. The system Cr—C has been the subject of many investigations [483-492].

One of the recent researches, in which the definitive phase diagram was established, is [492]. According to the diagram proposed in it (Fig. 54), the system Cr—C contains three carbides: Cr_3C_2, Cr_7C_3, and $Cr_{23}C_6$, which melt incongruently.

On rapid cooling the metastable system Cr—Cr_7C_3 is formed, and on slow cooling the carbide $Cr_{23}C_6$. This indicates that the coordination of the chromium and carbon atoms in the melt is closer to the coordination in Cr_7C_3 than to that in $Cr_{23}C_6$. Attempts to obtain the carbide CrC [492], whose existence had been reported in [483], did not meet with success.

The solubility of carbon in chromium has been studied in [371, 493-495].

The solubility of carbon in chromium, which reaches a maximum of 0.3-0.4% C at about 1500°C, was confirmed in all the researches.

The electrical properties of chromium carbides have been studied in [518, 843], where it was shown that the Hall coefficient and the specific electrical resistivity vary monotonically with rise in carbon content.

The small values of the thermo-emf coefficients, the positive values of the temperature coefficients of the electrical resistivity, and the comparatively high resistance provide evidence that all three carbides have metallic conduction.

In the lowest carbon carbide, $Cr_{23}C_6$, conduction is predominantly of the hole type. With increase in carbon content the nature of the conduction in the carbides changes and becomes mainly electron type.

Measurements of the temperature coefficient of electrical resistivity have shown [902] that in the lowest carbon carbide,

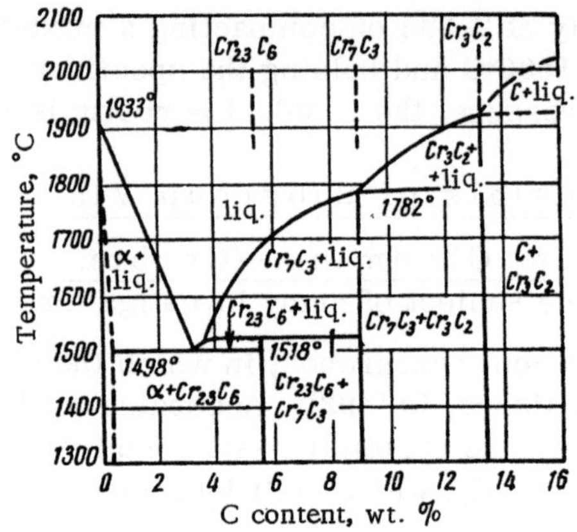

Fig. 54. Phase diagram of the system Cr – C [492].

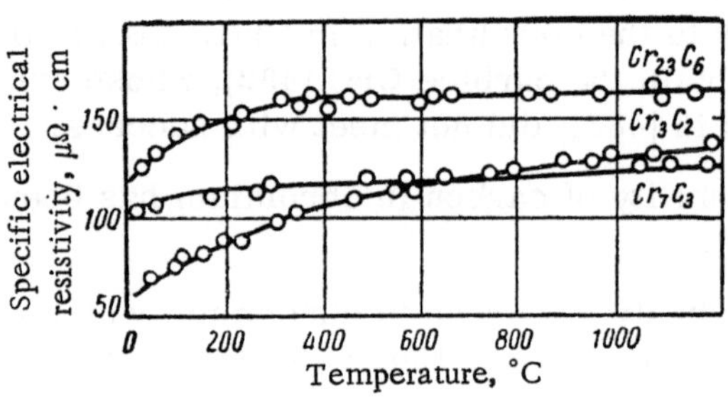

Fig. 55. Temperature dependence of the specific electrical resistivity of chromium carbides.

$Cr_{23}C_6$, the resistivity is practically constant over the temperature range 350–1200°C. The temperature dependence of the resistivity increases with rise in carbon content in the phase (Fig. 55).

A study of the evaporation of chromium carbides has shown [244, 866] that it proceeds in the following stages:

$$^1/_9\, Cr_{23}C_{6\,\text{solid}} \rightleftarrows {}^2/_9\, Cr_7C_{3\,\text{solid}} + Cr_{\text{gas}},$$

$$^2/_5\, Cr_7C_{3\,\text{solid}} \rightleftarrows {}^3/_5\, Cr_3C_{2\,\text{solid}} + Cr_{\text{gas}},$$

$$^1/_3\, Cr_3C_{2\,\text{solid}} \rightleftarrows {}^2/_3\, C_{\text{gas}} + Cr_{\text{gas}}.$$

The thermodynamic characteristics of chromium carbides are given in [23, 173, 316, 368, 496–500].

The most widely used method of producing chromium carbides is by the reaction of the oxide with carbon black in accordance with the equations

$$3Cr_2O_3 + 13C = 2Cr_3C_2 + 9CO,$$

$$7Cr_2O_3 + 27C = 2Cr_7C_3 + 21CO,$$

$$23Cr_2O_3 + 81C = 2Cr_{23}C_6 + 69CO.$$

The equilibrium of the system $Cr_2O_3 - C - CO - Cr_xC_y$ was first studied in [233].

In the production of hard alloys based on Cr_3C_2 [11, 22], briquets of a mixture consisting of 74% Cr_2O_3 and 26% carbon black are heated in a hydrogen atmosphere at 1600°C in an electric resistance furnace. This method leads to the formation of products of mixed composition containing up to 30% of the lower carbides.

A study of the conditions for preparing the carbide Cr_3C_2 by the method described in [501] has shown that a carbide similar in composition to Cr_3C_2 is obtained in the temperature range 1400-1600°C. At 1400°C a carbide containing a considerable amount of impurities in the form of the lower carbides is formed, while at temperatures above 1600°C the carbide dissociates and becomes decarburized.

The production of chromium carbides by the reduction of the oxide with carbon has been studied in [502-507, 822].

A detailed study of the conditions and mechanism by which the single-phase carbide Cr_3C_2 is obtained has shown [505, 822] that formation of the carbide begins at 1150-1200°C and proceeds in stages via the formation of the lower carbides. At 1500-1600°C the single-phase carbide Cr_3C_2 containing practically no free carbon is formed. Raising the temperature above 1600°C leads to the formation of products containing considerable amounts of the lower carbide Cr_7C_3 and free carbon. This phenomenon of decarburization at high temperatures is due to the fact that on raising the temperature above 1600°C the rate of graphitization of the carbon black increases at the same time as the rate of carbide formation increases; as a result, the vapor tension of acetylene at a given temperature falls below the equilibrium value corresponding to the calculated content of combined carbon in Cr_3C_2. Partial decarburization takes place, and part of the Cr_3C_2 is converted into the carbide Cr_7C_3, which has a lower carbon content.

By investigating the effect of temperature, reduction time, and composition of the mixture, it has been possible to determine more closely the conditions for obtaining the single-phase carbide Cr_3C_2: briquets made from a mixture of the calculated composition with addition of a 5% solution of dextrin are heated in an electric resistance furnace with graphite heaters by gradually raising the temperature to 1500°C over a period of 30–40 min and maintaining them at this temperature for 1.5–2 h.

An original method for producing porous parts from the carbide Cr_3C_2 has been developed in [938]. In this method the specimens are obtained directly during formation of the carbide. The particular conditions under which the chromium oxide reacts with carbon — the use of a solution of polyvinyl alcohol in water as a plasticizer, the conditions in which the charge is mixed and dried, and the slow raising of the temperature during carburization — create circumstances in which the carbon monoxide liberated during the reduction results in a highly porous specimen, while the carbide particles have a high activity at the moment when they are formed and sintering proceeds simultaneously in the porous specimen. The porosity of the specimens reaches 70%.

Porous chromium carbide specimens can also be produced by the introduction, as activating additions, of chlorides: 2–4% $CrCl_3 \cdot 10H_2O$, 6–8% $CoCl_2 \cdot 6H_2O$, and 1–2% $NiCl_2 \cdot 6H_2O$.

The activating effect of these additions permits the sintering temperature to be reduced by 200–300°, while maintaining the high strength of the specimens and a porosity of 50% [942].

An investigation into the conditions for obtaining the carbide Cr_7C_3 by reducing chromium oxide with carbon black in accordance with the equation

$$7Cr_2O_3 + 27C = 2Cr_7C_3 + 21CO$$

has shown [505, 506, 822] that the optimum temperature is 1200–1300°C and the composition of the charge containing 97–98% carbon black of the calculated composition.

It is convenient to produce the carbide Cr_7C_3 in two stages: briquets of the charge are heated at 1200-1300°C for 1-1.5 h; they are then ground and again heated at the same temperature for 20-30 min.

The production of the single-phase lower carbon carbide $Cr_{23}C_6$ [507] by the reduction of Cr_2O_3 is difficult, since it is prevented by the formation of the thermodynamically more stable higher carbides.

Investigations into the kinetics and mechanism of reduction of chromium oxide in vacuum have shown [41, 508, 865] that the reaction proceeds on account of the carbon monoxide evolved and is autocatalytic in character — the reduction rate increases rapidly in the initial stage, reaches a maximum, and then slowly decreases. From this it has been concluded that in the initial stage of the reduction of chromium oxide the limiting factor is the crystallochemical transformation of the oxide into metallic chromium or carbide. Then the part played by gasification of the carbon increases, and the reaction is sensitive to all the factors that affect the rate of regeneration of CO. Toward the end of the reduction process, when the surface of the reaction zone diminishes, the role of this factor again decreases.

In [497] all three chromium carbides have been obtained by the reaction of chromium oxide with carbon black in vacuum. The carbide Cr_3C_2 was obtained by heating a mixture of Cr_2O_3 and carbon black in the molar proportions 3 :13 at 1150-1200°C; Cr_7C_3 by heating a mixture of Cr_2O_3 and carbon in the molar proportions 7 :27 at 1200-1250°C; and $Cr_{23}C_6$ by heating Cr_2O_3 and carbon in the molar proportions 2 :7 at 1250-1300°C.

More recent researches into the conditions for obtaining chromium carbides in vacuum have shown that Cr_3C_2 close to the calculated composition is formed at 1400-1500°C; at 1150-1200°C products consisting of a mixture of all the carbides and of chromium oxide are formed.

The production of chromium carbides in vacuum requires the use of more complicated apparatus than does their production in a hydrogen atmosphere and the process takes longer; consequently it is not suitable for industrial use.

Chromium carbides have been obtained by synthesis from the elements in [25, 483, 487, 488].

It has not been possible to obtain the pure single-phase carbide $Cr_{23}C_6$ by synthesis from the elements or by the reduction of chromium oxide by carbon black. However, this carbide, close to the calculated composition and containing no higher carbides and oxides, has been obtained by the reaction of the higher carbide with chromium oxide in accordance with the equation $Cr_2O_3 + Cr_3C_2 \rightarrow Cr_{23}C_6 + CO$ in vacuum at 1200°C.

In [50] the carbide Cr_3C_2 has been obtained by the carburization of chromium by methane in the presence of hydrogen at 600-800°C, in accordance with the reaction

$$Cr + H_2 + CH_4 \rightarrow Cr_3C_2 + H_2 + (CH),$$

where the symbol (CH) denotes the dissociation products of hydrocarbons.

Very hard and wear-resistant coatings, consisting mainly of Cr_3C_2, have been produced on iron surfaces by the decomposition of chromium carbonyl in the presence of hydrogen at temperatures from 250 to 850°C [509].

In [510], chromium carbide coatings were produced on specimens of Armco iron, steel 18KhNMA, and steel KhG by electrolytic chromium plating in a bath containing 150 g CrO_3 and 1.5 g H_2SO_4 per liter of water at a current density of 35 A/dm^2, and electrolyte temperature of 50°C, and a plating time of 3 h. The chromium-plated specimens were then carburized in a vertical tubular furnace by passing through it a mixture of benzene vapor and hydrogen at 950 and 1050°C for 3 and 8 h.

The coatings consist of an outer thin layer of the carbide Cr_3C_2, an intermediate thick layer of Cr_7C_3, and a comparative thin layer of $Cr_{23}C_6$. Beneath this is a layer of residual uncarburized chromium.

Subsequently, a study was made of the production and properties of coatings formed by the gas carburization of chromium electrodeposited on steel in a stream of benzene vapor mixed with nitrogen [511].

Metallographic and x-ray phase analyses have revealed the existence of three phases arranged in layers: an outer thin layer of the higher carbide Cr_3C_2, an intermediate thick layer of Cr_7C_3, and an inner thin layer of the nitride Cr_2N.

Reaction diffusion in the system Cr — N — C takes place by the diffusion of nitrogen and carbon atoms through the layer formed to the metal. As the atomic radius of nitrogen is less than that of carbon, the diffusion mobility of nitrogen atoms is higher than that of carbon atoms, and nitrogen penetrates more rapidly into the depth of the metal.

A wear-resistant layer of chromium carbide has been formed on the surface of steel parts by the deposition of chromium from gaseous compounds followed by its reaction with the carbon present in the surface layer of the parts.

Numerous investigations have been concerned with the separation of chromium carbides from ferroalloys and steels by utilizing the difference in solubility between the carbides and the metallic constituents. It was found in [513] that on the electrolytic solution of chromium steel in hydrochloric acid of density 1.02, the insoluble residue consists of a mixture of iron and chromium carbides.

The carbides Cr_7C_3 and $Cr_{23}C_6$ have been separated from chromium steels [515], Cr_7C_3 from chromium — manganese steels [516], and $Cr_{23}C_6$ from chromium — nickel steels [517].

Although all these methods are of considerable theoretical interest, they are not used on an industrial scale.

By investigating the conditions for preparing dense specimens by hot compaction [130], it has been possible to obtain specimens of density 99-99.5%, using a temperature of 1750°C and a specific compaction pressure of 1500 MN/m^2 (150 kgf/cm^2). In this process the content of the lower carbon carbide Cr_7C_3 increases from 5% in the original powder to 15%.

<u>The System Molybdenum — Carbon.</u> A compound containing 5.8% C was first obtained in [522], and it was given the formula Mo_2C. Later it was shown on the basis of an x-ray analysis of Mo — C alloys [522] that at about 30 at.% C a solid solution of carbon in molybdenum is formed, having a close-packed hex-

agonal lattice. The lattice parameter *a* increases from 0.2993 to 0.3004 nm with rise in carbon content from 30 to 39 at.% C.

However, subsequent investigations do not support this conclusion. According to the results of a number of authors [523–525], the solubility of carbon in molybdenum is negligible.

The following data on the solubility of carbon in molybdenum are based on a metallographic investigation [525]:

Temperature, °C	1650	1925	2200
Solubility, wt.%	0.005-0.009	0.012-0.013	0.018-0.022

From this it may be concluded that Mo_2C is a chemical compound, the homogeneity range extending from 5.4 to 5.9% C at 1400–2200°C.

By treating metallic molybdenum with a mixture of CO and H_2, a product has been obtained [526] having an fcc lattice with parameter 0.414 nm and containing 30–35 at.% C. The formula Mo_2C has been assigned to this carbide, it being assumed that there is a low-temperature cubic modification and a high-temperature hexagonal one.

In a study of the crystal structure of molybdenum carbides [11], new lines were found on the x-ray diffraction diagrams of Mo_2C specimens heated to temperatures above 1400°C. It was concluded that at high temperatures a new phase β-Mo_2C is formed which has the same crystal structure as the low-temperature phase α-Mo_2C with the same parameter *a* and a larger parameter *c*.

The existence of a rhombic modification of Mo_2C with lattice parameters $a = 0.4735$, $b = 0.6038$, and $c = 0.5208$ nm, has been established in [950].

However, the existence of the high-temperature modifications of Mo_2C described has not been confirmed in subsequent investigations.

On alloying molybdenum with aluminum and carbon in a carbon crucible at the boiling point of aluminum, Moissan obtained after cooling, in addition to aluminum carbide and graphite, crystals of a compound containing 11.1% C and corresponding to the formula MoC [527]. The existence of this compound has also been reported in [189, 528–530]. It was shown in [530] that MoC can be stabilized in WC — MoC and NbC — MoC solid solutions, but it cannot be obtained by synthesis from the elements or by the reaction of molybdenum with carbon-containing gases.

The possibility of obtaining molybdenum carbides by the reaction of the metal with methane has been studied in [51]. In this work there was obtained, in addition to the carbide Mo_2C, a carbide richer in carbon, MoC, which is metastable and which decomposes at high temperatures into Mo_2C and carbon. No data were given on the structure of the monocarbide.

An x-ray study of molybdenum carbide specimens containing about 7.8% of combined carbon showed them to consist of a mixture of 60% Mo_2C and 40% MoC [531]. In addition, certain lines of an fcc lattice with a parameter of 0.419 nm were observed on the x-ray diffraction diagrams, but the authors did not succeed in interpreting them.

Three phases were found in an investigation of molybdenum–carbon alloys by Sykes [532]; the α-phase, a solid solution of carbon in molybdenum (the limiting solubility being 0.09% at 1500-2000°C); the β-phase, containing 5.4-6.0% C at 1400-2200°C; and the γ-phase, formed at 2100°C and containing 12.3-13.0% C. However, according to an x-ray analysis, this phase differs somewhat from the monocarbide MoC previously discovered.

The α- and β-phases form a eutectic at 2200°C, 1.5% C.

The phase diagram has been made more accurate by melting-point measurements on alloys containing 40-75 at.% C [533]. The MoC — C eutectic point corresponding to 70 at.% C (22.5 wt.%) has been found to lie at 2400°C (Fig. 56).

Later investigations [534] have confirmed the existence of the two carbides, β-phase (Mo_2C) and γ-phase (MoC), and have also revealed a new phase (γ') formed by the combination of the other two.

Carburization of molybdenum at low temperatures leads to the formation of the γ'-phase after the β-phase. On heating the γ'-phase above 800°C in a CO atmosphere or in vacuum, it is converted into the γ-phase, which is more stable than the γ'-phase at high temperatures. At low temperatures, the γ-phase is less stable, and the rate of transformation of γ' into γ is very small.

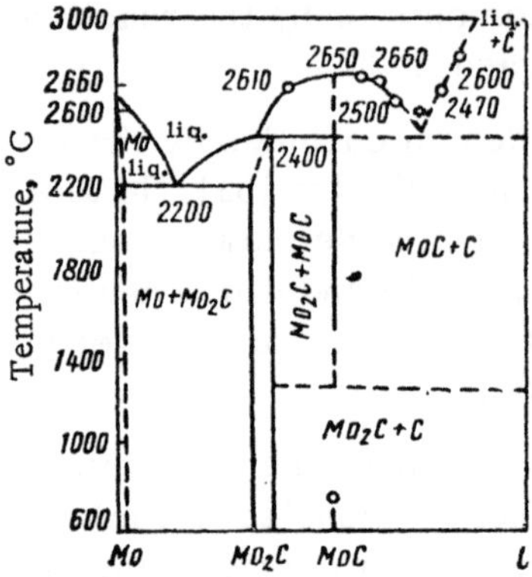

Fig. 56. Phase diagram of the system Mo — C.

The γ'-phase has a hexagonal lattice containing four molybdenum atoms in the unit cell, which is similar to the W_2B_5 cell.

Table 25 gives the principal x-ray parameters of all three phases; these are in good agreement with the data contained in [522, 523, 528, 548, 549].

From the arrangement of the molybdenum atoms, the γ'-phase appears to be intermediate between the β- and γ-phases. It also is represented by the formula MoC. Data are available on the existence of α-MoC, with an fcc lattice and parameter a = 0.427 nm [398].

An x-ray and metallographic study of the system V — Mo — C has shown [535] that the phases Mo_2C and η-Mo_3C_2, which crystallize in a hexagonal lattice, exist in the Mo — C system. On rapid cooling the carbide α-Mo_3C_2 is obtained; it crystallizes in a cubic lattice with parameter a = 0.4281 nm. This phase is stable at 1800°C. On slow cooling and annealing it is converted into the η-phase and breaks down into Mo_2C and C. The η-phase can be stabilized by adding up to 2% B and 1% U or Th to the Mo — C alloys.

The existence of the carbide η-Mo_3C_2 has been confirmed in [536, 537].

The carbide Mo_2C (calculated carbon content = 5.91%) is a dark-gray metallic powder.

The characteristic temperatures θ_a and θ_c and the thermal coefficients of linear expansion of Mo_2C in the directions of the a and c axes have been calculated from x-ray data [553]. It was shown that in the temperature range 17-270°C, $\alpha_c > \alpha_a$ and α_a = $6.9 \pm 0.6 \cdot 10^{-6}$ deg^{-1} and $\alpha_c = 8.9 \pm 0.4 \cdot 10^{-6}$ deg^{-1}. No difference was found between the characteristic temperatures: $(m\theta_a^2)$ = $(m\theta_c^2) = 21.3 \pm 4.0 \cdot 10^{-18}$ g · deg^2.

TABLE 25. X-Ray Parameters of Phases in the System Mo — C

Parameters	Phase		
	β	γ′	γ
a, nm	0 3002	0.2932	0.2898
c, nm	0.4729	1.097	0.2809
c/a	1.574	3.742	0.969
No. of Mo atoms per unit cell	2	4	4
Volume of cell per atom of Mo, nm^3	0.01844	0.02042	0.02042

TABLE 26. Reduction of MoO_3 by Carbon Black in Various Gaseous Atmospheres (Temperature 950°C, Time of Heating 0.5 h)

Atmosphere	Content, %			Completeness of reaction, %
	C_{total}	C_{free}	C_{comb}*	
Vacuum	5.6	1.2	4.45	78.5
Hydrogen + 1.3% propane	5.8	0.73	5.17	87.5
Hydrogen + hydrogen chloride (6 : 1.3) + 1.3% propane	6.2	0.46	5.77	97.5
Hydrogen + hydrogen bromide (6 : 1.3) + 1.3% propane	6.0	0.3	5.72	97.0
Hydrogen + hydrogen iodide (6 : 1.3) + 1.3% propane	5.9	0.15	5.75	98.0

* $C_{comb.}$ is calculated as $[(C_{total} - C_{free})/(100 - C_{free})] \cdot 100$.

The thermodynamic properties of Mo_2C are given in [60, 65, 135, 550, 551].

The reduction of molybdenum anhydride by carbon in a hydrogen atmosphere, using hydrogen halide compounds as catalysts, has been investigated in [298].

The experimental results given in Table 26 indicate that in the presence of the hydrogen halide compounds products are formed whose composition is close to the calculated composition of Mo_2C, and the reduction process goes practically to completion.

The carbide Mo_2C has also been obtained by the reduction of molybdenum glance, calcium molybdate, by calcium carbide [485].

A detailed investigation of the mechanism of the reaction of molybdenum anhydride with carbon black has been made in [541]. It was found that the reduction of the anhydride in an argon or CO atmosphere proceeds by the following stages:

1. Reduction of MoO_3 to MoO_2 at 600–700°C:

$$2MoO_3 + C = 2MoO_2 + CO_2.$$

2. Reduction of MoO_2 to molybdenum at 900–1000°C:

$$MoO_2 + C = Mo + CO_2,$$
$$MoO_2 + 2CO = Mo + 2CO_2,$$
$$CO_2 + C = 2CO.$$

3. Formation of Mo_2C by the reaction

$$2Mo + C = Mo_2C.$$

No intermediate oxides between MoO_3 and MoO_2 and Mo are formed.

The industrial production of molybdenum carbide [22] consists in reducing the anhydride with hydrogen at 900°C, followed by synthesis of the carbide from the metallic powder and carbon in a hydrogen atmosphere at 1400–1500°C.

Molybdenum carbide was first obtained by synthesis from the elements in [25, 527].

A fairly widely used method of producing molybdenum carbide is by deposition from the gaseous phase. Mo_2C has been obtained by the carburization of molybdenum powder by means of CO at 600–1000°C [543].

A study of the equilibrium of the reaction $2Mo + CH_4 = Mo_2C + 2H_2$ has shown [51, 544] that at 700–850°C quantitative decomposition of methane takes place with the formation of the carbide Mo_2C containing 5.8–6.0% C.

Molybdenum carbide has also been obtained on heating a molybdenum wire by passing a current through it in an atmosphere of naphthalene [545], using a mixture of hydrogen and molybdenum carbonyl vapor, which decomposes at a temperature of 300–800°C and a pressure of 13.33–39.99 N/m^2 (0.1–3 mm Hg). In this process either Mo_2C or a mixture of Mo_2C, MoC, and Mo is obtained, depending on the experimental conditions.

The electrolysis of molten salts is of considerable interest as a method of producing molybdenum carbides. It was found in [63, 546] that on electrolyzing at 700°C mixtures of molten sodium carbonate, sodium metaborate, lithium fluoride, and molybdenum trioxide in various proportions, Mo_2C, MoC, or a mixture of the two is deposited on the graphite electrode.

The controlling factor in this process is the ratio of MoO_3 to CO_2. At $MoO_3 : CO_2$ from 1 : 5.5 to 1 : 7, Mo_2C containing 5.9% C and 94.0% Mo is deposited.

On using a bath composition with a ratio of MoO_3 to CO_2 of 1.21-1.28 the monocarbide MoC containing 11.5% C and 87.5% Mo is deposited. At a ratio of MoO_3 to CO_2 of 1.5-1 : 18, a mixture of the two carbides may be deposited.

In the electrolytic decomposition of molybdenum — carbon alloys in dilute hydrochloric acid (density 1.94) at a current density of 0.3-0.4 A/cm^2, Mo_2C is left in the residue [523]. The carbide Mo_2C can also be deposited electrolytically from melts of Ni_3Mo_3C — Co_3Mo_3C — Fe_3Mo_3C, which are very unstable and which on cooling decompose with the formation of Mo_2C [547].

The System Tungsten—Carbon. In an investigation of the W — C system by chemical and metallographic methods [24], the existence of a carbide corresponding to the formula W_3C was suggested, in addition to W_2C and WC; however, its existence was not subsequently confirmed.

As a result of an x-ray analysis and a study of the physical properties of W — C alloys, a new phase was found with a structure similar to that of W_2C [430]. This phase is regarded as a polymorphic modification of β-W_2C. It is unstable and is easily converted into α-W_2C.

It has been established in [950] that there is a modification of W_2C which crystallizes in a rhombic lattice and which differs from the pseudohexagonal structure of Mo_2C by the distorted stacking of the metal atoms. The lattice parameters of this phase are: a = 0.4721, b = 0.6030, and c = 0.5180 nm.

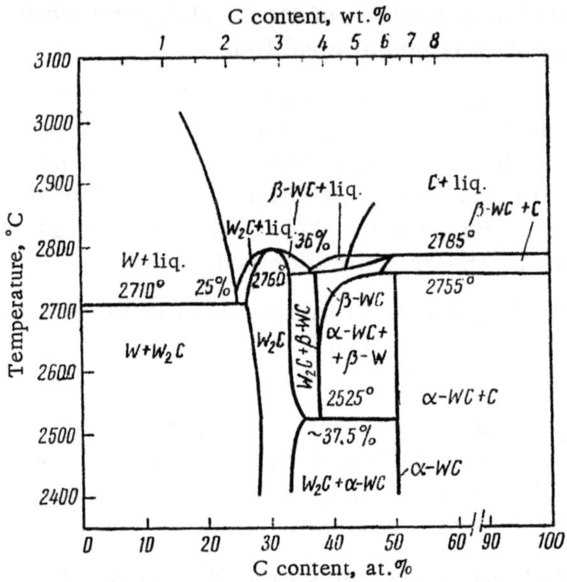

Fig. 57. Phase diagram of the system W–C [860].

Numerous investigations of the system W — C [51, 294, 306, 554, 859, p. 250] have led to the conclusion that the system contains two carbides: W_2C with a hexagonal close-packed lattice and the monocarbide WC with a simple hexagonal lattice [259, 263, 522, 571]. The monocarbide is the most stable compound in the system W — C [51]. The maximum solubility of carbon in tungsten is 0.3 at.% at 2400°C [306], and it decreases rapidly with falling temperature, being 0.05% at 2000°C and only traces at lower temperatures.

The tungsten-rich solid solution and the lower carbon carbide W_2C form a eutectic containing 2.0 wt.% C and melting at 2710°C.

At 3.5 wt.% C the eutectic W_2C + WC is formed, which melts at 2760°C.

In addition, a second modification of the carbide β-WC has been found; it crystallizes in a cubic lattice and is stable at temperatures above 2525°C [859, p. 251].

A phase diagram of the system W — C has been constructed on the basis of all the available information [859, p. 251] and is shown in Fig. 57.

The characteristic temperatures and coefficients of thermal expansion along the *a* and *c* axes have been calculated from the x-ray data in [553]. It is shown that in the temperature range 22-400°C for WC, in which the ratio *c*/*a* is close to unity, $m\theta_a^2 = m\theta_c^2 = 32.4 \pm 4.6 \cdot 10^{-18}$ deg, and the coefficients of linear expansion $\alpha_a = 3.84 \cdot 10^{-6}$ deg^{-1}, $\alpha_c = 3.90 \cdot 10^{-6}$ deg^{-1}; for W_2C in the temperature range 12–400°C, $m\theta_a^2 = 29.6 \pm 6 \cdot 10^{-18}$ deg; $m\theta_c^2 = 19.3 \pm 2.9 \cdot 10^{-18}$ deg; $\alpha_a = 6.4 \pm 0.16 \cdot 10^{-6}$ deg^{-1}; $\alpha_c = 8.1 \pm 0.22 \cdot 10^{-6}$ deg^{-1}.

The effects of high pressures and temperatures on the properties of tungsten carbide have been studied in [575]. The specimens were obtained by compacting WC powder containing 6.08% C and sintering in a hydrogen atmosphere at 2100°C. The sintered specimens were subjected to hydrostatic compression up to 10,130 MN/m^2 (100,000 atm) while being heated to 2100°C. Some of the specimens were annealed at 1500°C for 1.5 h.

The microhardness of specimens subjected to hydrostatic compression but not annealed was 27,400–31,360 MN/m^2 (2800-3200 kgf/mm^2); after annealing it decreased to a value corresponding to the hardness of specimens sintered in the ordinary way without compression.

In studying the adsorption, migration, and evaporation of barium on carburized tungsten [576], a value of $3.6 \cdot 10^{-19}$ J (2.25 eV) was found for the work function of W_2C covered with barium for a coating corresponding to maximum emission current.

The thermodynamic properties of tungsten carbide are described in [60, 135, 542, 573, 574].

Tungsten carbide was first obtained by melting tungsten in an electric arc furnace [24, 49, 292, 430]. This method, by which it is possible to obtain the lower carbon carbide W_2C close to the calculated composition, has great industrial importance.

Two principal methods of producing cast tungsten carbide alloys are known [556].

In the first, tungsten produced by the reduction of tungsten anhydride with carbon is carburized and melted in an electric resistance furnace with graphite heaters at 2700°C. The melt is poured into a graphite mold by tilting the furnace.

In the second method, tungsten produced by the reduction of tungsten anhydride with hydrogen is carburized and melted in a graphite crucible at 3000°C. The melt is poured into special molds by means of a centrifugal device.

The higher carbide WC of exact composition cannot be produced by this method because it melts peritectically to give W_2C and graphite.

At the present time, the most widely used method of producing tungsten carbide is by synthesis from the elements in a hydrogen atmosphere.

The reactions $W + C = WC$ and $2W + C = W_2C$ were first thoroughly investigated under various conditions in [49, 430, 557].

In a hydrogen atmosphere in which hydrocarbons are formed, these create a carburizing atmosphere favorable to the production of carbides. This was confirmed in later investigations, which showed [285] that in a hydrogen atmosphere acetylene is formed by the reaction of hydrogen with the carbon black and the tube walls and boat. The vapor tension of acetylene is less over the tungsten particles than it is over the carbon particles, and as a result carbon atoms are transferred from the carbon particles to the tungsten particles. In a CO atmosphere the transfer via the gaseous phase takes place in accordance with the reaction $C + CO_2 = 2CO$, the equilibrium of which is completely displaced in the direction of CO formation at temperatures above 800°C.

An investigation of the reaction of metallic tungsten powder with a mixture of carbon monoxide and hydrogen in the temperature range 1000-1100°C at $CO : H_2$ ratios of 0.5-1 has shown [834] that the carbide WC is formed in this way though it is not possible to achieve complete transformation of the tungsten into the carbide.

As a result of an investigation of the parameters of the reaction diffusion of carbon into tungsten [382], a value of the activation energy of 14.24 kJ/mole (34,000 cal/mole) was obtained, and $D = 1.6 \cdot 10^{-4} \exp(17{,}000/T)$.

The mechanism of carburization of tungsten from the solid and gaseous phases has been studied in detail in [558]. As a result, specimens were obtained which were covered with a diffusion layer consisting of two parts: WC outside and W_2C inside.

An investigation of the reaction diffusion of carbon into tungsten in an atmosphere consisting of a mixture of benzene vapor and hydrogen at temperatures up to 1200°C has shown [559] that appreciable reaction between tungsten and carbon begins at 1000°C and that the process obeys a parabolic law.

Metallographic and x-ray analyses of the specimens have confirmed the presence of two phases in the diffusion zone; WC outside and W_2C inside. A study of reaction diffusion in the ternary system W — N — C has shown that this process also obeys the parabolic law. The diffusion zone is similar in composition to that formed by the diffusion of carbon into tungsten. Nitrogen retards the diffusion of carbon into tungsten. Obviously, the carbides formed in the diffusion zone contain a certain amount of dissolved nitrogen, and the diffusion of carbon through the carbide phases containing the dissolved nitrogen is inhibited. The formation in the given case of phases with the crystal lattices of the carbides indicates that at the experimental temperatures used the thermodynamic stability of the carbides is greater than that of the nitrides.

The method of producing the monocarbide by the reaction between tungsten powder and carbon black in a hydrogen atmosphere is widely used in the hard-alloy industry. The process is carried out in electric resistance furnaces with graphite heaters. In [560] a furnace for the carburization of tungsten has been suggested which has rectangular Alundum muffles and molybdenum heaters with a working temperature of 1450°C. This furnace is characterized by very stable thermal conditions which ensure a high quality in the finished product.

Metallic tungsten powder is obtained from tungsten anhydride, tungstic acid, or ammonium paratungstate [561]. The tungsten powder is mixed with 6.3-6.8% of carbon black, and the resulting mixture is heated in the form of powder or briquets at 1400-1600°C [22].

In [597] an x-ray and electron-diffraction study has been carried out on carbide powders synthesized from tungsten obtained by the reduction of tungstic acid and paratungstate. The carbide powder is more finely divided in the first case than in the second. However, after grinding for 100 h, the average particle size is the same in both cases, being 0.073 μm, with a distortion of the crystal lattice of 0.16%. Increasing the grinding time to 300 h has practically no effect on the particle size, but the lattice distortion reaches 5%.

In contrast to all other refractory compounds, the lower carbide W_2C is produced on a commercial scale by melting. A mixture of commercially pure tungsten with 3% of carbon black and iron, in the form of briquets, is melted at 3000–3250°C in tubular graphite furnaces of special construction or in high-frequency furnaces.

The conditions for obtaining tungsten carbide by the direct reduction of the anhydride with carbon black have been investigated in [562]. The reduction proceeds efficiently in a hydrogen atmosphere; in this process the carbon content in the charge is 12.5% instead of 20.6% as required theoretically. This is due to the participation in the reduction of acetylene formed in the reaction zone.

The effect of additions of various hydrogen halides on the reaction of tungsten anhydride with carbon (Table 27) has been investigated in [298]. The addition of the hydrogen halide compounds to the gaseous mixture considerably increases the combined carbon content and the degree of completeness of the reaction and reduces the free carbon content. The most effective addition is hydrogen iodide.

The decarburization of tungsten carbide by moist hydrogen has been studied over the temperature range 800–1000°C in [563]. According to thermodynamic calculations, the change in free energy in the decarburization of tungsten carbide by dry hydrogen has a negative value up to 227°C (Table 28). Consequently, at the temperatures at which the carbide is formed, dry hydrogen cannot decarburize it. In using moist hydrogen, decarburization takes place on account of the moisture at temperatures above 1230°C, i.e., at temperatures at which the carbide is formed.

Investigations have shown that when using moist hydrogen, a sudden change in the rate of decarburization takes place at 950°C (Fig. 58).

Thus, if the hydrogen contains water vapor, there is considerable decarburization of the carbide, which increases with temperature.

TABLE 27. Effect of Composition of the Gaseous Atmosphere on the Composition of Tungsten Carbide (Ratio of WO_3 : C in the Charge = 3.1 : 1; Temperature = 1100°C; Time of Heating = 1 h)

Atmosphere	Content, %			Completeness of reaction, %
	C_{total}	C_{free}	C_{comb}*	
Hydrogen + 0.5% propane	6.0	1.85	4.23	69.2
Hydrogen + HCl (6 : 1.3) + 0.5% propane	6.3	0.66	5.69	92.0
Hydrogen + HBr (6 : 1.3) + 0.5% propane	6.2	0.45	5.77	92.5
Hydrogen + HI (6 : 1.3) + 0.5% propane	6.05	0.1	5.95	98.0
Vacuum	5.8	2.3	3.58	60.4

*C_{comb} is calculated as $[(C_{total} - C_{free})/(100 - C_{free})] \cdot 100$.

TABLE 28. Values of the Loss of Free Energy in Decarburization of the Carbide WC

Reaction	ΔF							
	27° C		227° C		730° C		1230° C	
	kJ/mole	kcal/mole	kJ/mole	kcal/mole	kJ/mole	kcal/mole	kJ/mole	kcal/mole
$WC + 2H_2 = W + CH_4$	−17.6	− 4.2	− 1.50	− 0.36	+48.8	+11.65	+107.3	+25.6
$WC + H_2O = W + CO + H_2$	+82.1	+19.6	+95.1	+22.7	+21.4	+ 5.1	− 50.7	−12.1

In studying the mechanism of reduction of tungsten anhydride by carbon, it has been found [564] that the reduction of WO_3 by carbon black depends on temperature and passes through the stages of formation of the lower oxides and α-tungsten. The whole process follows the series of equations

$$WO_{3\,solid} \rightarrow WO_{3\,gas},$$

$$WO_{3\,gas} + C_{solid} \rightarrow W_{solid} + CO_{gas} + CO_{2\,gas},$$

$$W_{solid} + WO_{3\,gas} \rightarrow WO_{2\,solid + gas},$$

$$WO_{2\,gas} + C_{solid} \rightarrow W_{solid} + CO_{gas},$$

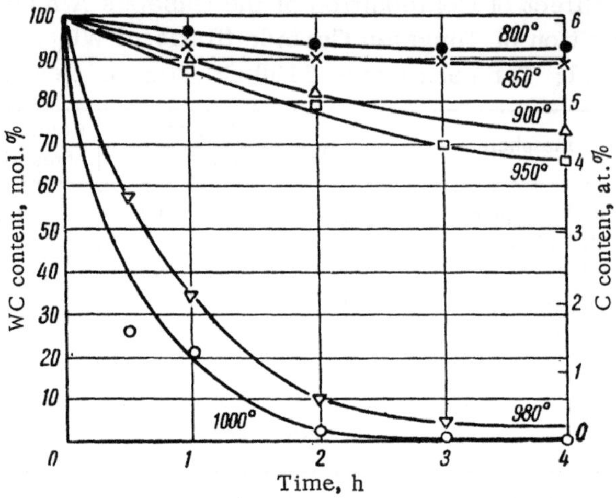

Fig. 58. Dependence of the rate of decarburization of WC on time and temperature.

$$CO_2 + C \rightarrow CO,$$

$$W_{solid} + C_{solid} \rightarrow WC_{solid} .$$

A method is also known by which tungsten carbides can be obtained by reduction of the tungstate by carbon at 1400°C [565]. The products obtained consist of a mixture of WC and W_2C_{1-x}.

Tungsten carbides are also obtained by deposition from the gaseous phase, the most widely used source being tungsten carbonyl.

Thus, W_2C_{1-x} has been obtained [526] by decomposition of the carbonyl vapor, while the monocarbide WC has been obtained by decomposing the carbonyl at 1000°C [566]. Carbides have also been obtained by heating a mixture of tungsten carbonyl and hydrogen at 300-800°C [50].

In [568] the carbides were obtained by subjecting the carbonyl to thermal dissociation at 650-1000°C and then heating the dissociation products at 1800-1850°C in a CO atmosphere.

In addition, tungsten carbides have been produced by heating a mixture of hydrogen and methane with tungstic acid, tungsten anhydride, or ammonium paratungstate at 850-1000°C [569].

In studying electron emission it has been found [570] that W_2C_{1-x} is formed by the reaction of tungsten with carbon resulting from the decomposition of hydrogen molecules.

The tungsten carbides produced by the electrolysis of molten salts have been thoroughly investigated in [63], in which a mixture of molten sodium borate, lithium fluoride, and tungsten anhydride was electrolyzed. The nature of the carbide formed was determined mainly by the ratio of the amounts of CO_2 to WO_3 in the electrolyte. At a $CO_2:WO_3$ ratio of 3.3-6, W_2C_{1-x} is deposited on the cathode; at a ratio of 12-16, WC is deposited, while when the ratio is above 16, no carbide is formed.

Carbides of Metals of Subgroup VIIa

The System Manganese — Carbon. As a result of investigations of the system Mn — C in [42, 85, 577], a phase diagram has been proposed which contains only one carbide Mn_3C. It forms a continuous series of solid solutions with manganese over a wide temperature range. Subsequent investigations have shown [578, 579] that this carbide is polymorphic and exists in two forms α - and β -Mn_3C; of these, the α -modification is isomorphous with cementite and forms a continuous series of solid solutions with it; the β -modification is a high-temperature form, the $\alpha \rightarrow \beta$ transformation occurring at 2230°C. However, it was shown in [580] that Mn_3C exists only in the temperature range 950-1050°C, and it is possible that the change in properties of the carbide at 1050-1100°C is due not to a polymorphic transformation, but to decomposition of the carbide.

In addition to this carbide, the following compounds have also been found in the system Mn — C [580-584]; Mn_7C_3 and Mn_4C, which are similar to the corresponding chromium carbides, and also Mn_7C_2 and Mn_5C_2.

The carbide Mn_7C_3 (8.56% C) is isomorphous with Cr_7C_3. The carbon atoms in Mn_7C_3 are joined to one another in zigzag chains parallel to the z axis. It has, however, been suggested that Mn_7C_3 has a rhombic lattice corresponding to the space group D_{2h}^{16} Pnma, with parameters $a = 0.4546$, $b = 0.6959$, and $c = 1.1979$ nm [857], but this has not yet been adequately confirmed.

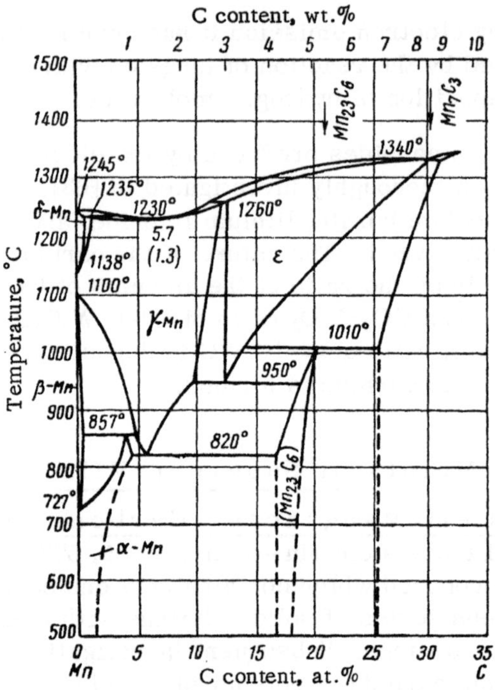

Fig. 59. Phase diagram of the system Mn — C [585].

According to x-ray studies [582], the composition of the carbide Mn_4C is more accurately represented by the formula $Mn_{23}C_6$. It is isomorphous with $Cr_{23}C_6$.

The carbide Mn_5C_2 is stable over a wide temperature range (up to 1050°C) and is identical in structure with the double carbide $(Mn, Fe)_5C_2$. The carbide Mn_7C_2 is stable over a narrow temperature range (850-1000°C); its structure is unknown and the formula is approximate [580]. The phase diagram of the system Mn — C put forward in [585] is shown in Fig. 59.

In later investigations of the system Mn — C [782, 856, 927, 946], it has been found that the carbon content in Mn_7C_2 is 5.48%; it crystallizes in a hexagonal lattice with parameters a = 0.7492 nm, c = 1.2070 nm, c/a = 1.61; its density is 7.30 ± 0.05 g/cm^3. These data point to the existence of a carbide phase whose composition is represented by the formula $Mn_{15}C_4$ and not Mn_7C_2.

The phase diagram of the system Mn — C has been studied in great detail in [584, 586]. Investigations of specimens containing 0-8.54% C, prepared in a high-frequency induction furnace from electrolytic manganese (99.8% Mn) and electrode graphite, have shown that carbon forms solid solutions with manganese. A high solubility of carbon is found in γ-Mn and a lower solubility in α-, β-, and δ-Mn. The δ- and β-solutions decompose peritectically at 1215 and 860°C. The γ-solution decomposes eutectically at 743°C to form the α-solution and Mn_4C. The carbides Mn_3C and Mn_4C are insoluble in one another, whereas Mn_7C_3 and Mn_3C form continuous series of solid solutions over a wide range of compositions. The author of [584, 586] gives no data on the carbides Mn_5C_2 and Mn_7C_2 which are mentioned in a number of other papers. This indicates that the definitive phase diagram of the system Mn — C has still to be published.

The properties of the carbides Mn_3C, Mn_7C_3, and $Mn_{23}C_6$ have been studied most thoroughly.

From an investigation of the thermodynamic properties of manganese carbides [173, 587, 589-596] it has been possible to derive equations for the temperature dependence of the change in free energy of formation of the various phases.

The results of studies by Knudsen's method of the vapor pressure of manganese over pure manganese and manganese carbide under conditions of equilibrium with graphite are given in [595]. These results show that the carbide Mn_7C_3 exists in equilibrium with graphite and manganese vapor in the chamber.

Manganese carbides are produced mainly by the carburization of solid manganese and by the reduction of Mn_3O_4 by carbon or calcium carbide.

To obtain the carbides by the reaction of solid manganese with carbon, various mixtures of manganese and carbon were heated in a graphite resistance furnace at 1250°C for 3 h in an argon atmosphere [588]. The raw materials used consisted of pure manganese, obtained by aluminothermy from Mn_3O_4 and purified by distillation at a temperature of 1200°C and a pressure of 1.33 N/m^2 (0.01 mm Hg), and graphite degassed by heating in vacuum. On heating the mixture in vacuum considerable quantities of manganese are lost as a result of its rapid evaporation above 1000°C.

The use of sugar carbon as a reducing agent is less effective than graphite. Thus at 1400°C only 50% of it reacts with the metal.

The carbide $Mn_{23}C_6$ has been prepared by sintering a metallic powder containing 99.77% Mn with carbon black at 1050°C in an argon atmosphere for several days [587]. The product contained admixtures of metallic manganese.

The System Rhenium — Carbon. Early investigations of the system Re — C [601] established that rhenium carbide is formed by the reaction of rhenium with carbon monoxide at 470–600°C. It decomposes at 1600°C. Later investigations [602, 603] showed that the phase diagram of the system Re — C is a simple eutectic one, the eutectic lying at 2480–2500°C, 16.9 at.% C.

The solubility of carbon in rhenium at 2500°C is 11.7 at.%, falling to 4.2% at 1800°C. No solubility of rhenium in carbon has been found. It has been shown in [603] that lines from an unknown phase occur on the x-ray diffraction diagrams of specimens containing 1.3–3.7 at.% C. The authors assume the formation of a carbide which is stable above 1900°C.

Investigation of the ternary system W — Re — C has revealed the existence of a ternary compound W_3Re_2C with a cubic structure [604]. Above 2500°C the carbide (W, Re)C is found, which has an fcc structure of the NaCl type.

In [932], a study has been made of the composition of the layers obtained on graphite and molybdenum substrates by the thermal decomposition of rhenium carbonyl in vacuum (1 mm Hg) at 875–925°K (on molybdenum) and 450–1000°K (on graphite). X-ray structural analysis of the layers revealed the presence on the diffraction diagrams, in addition to lines from metallic rhenium, of lines of a new phase which crystallizes in a hexagonal lattice of the Mo_2C type with lattice parameters $a = 3.02 \pm 0.01$, $c = 4.76 \pm 0.01$ Å, $c/a = 1.58$. The authors assume that the phase they obtained is similar to the Mo_2C phase in the Mo — C system and that it corresponds to the carbide Re_2C.

The production of the rhenium carbide phase requires further experimental confirmation.

The System Technetium—Carbon. This system has been investigated by the x-ray structural analysis of specimens obtained by heating a mixture of technetium with carbon black and also by heating technetium in an atmosphere of hydrogen and benzene vapor [605]. At 910°C, technetium dissolves up to 1% C. In this process the lattice parameters of hexagonal technetium increase to a = 0.2892 nm, c = 0.4470 nm, c/a = 1.590 (for pure technetium, a = 0.2740 nm, c = 0.4398 nm, c/a = 1.605).

At a content of 1.4-9% C, the products consist of a mixture of the metallic phase and the carbide TcC with an fcc lattice of parameter a = 0.3982 nm and a density of 11.5 g/cm^3.

Carbides of Metals of Group VIII

The System Iron—Carbon. The most complete reviews of the work carried out on the iron—carbon system are given in [85, 86, 606-612]. Investigations of the system [608-612] have established the existence in it, besides the stable carbide Fe_3C (cementite), of the unstable carbides Fe_2C and Fe_4C. Fe_2C exists in the form of the hexagonal ε-phase [613] and the ϰ-phase of another structure known as a Hagg carbide [611], which is more accurately represented by the formula $Fe_{20}C_9$. At 300-400°C, the hexagonal carbide is transformed into the Hagg carbide, which in turn changes into cementite [614, 615].

The instability of the ε-carbide is due to the fact that the interatomic distance Fe—Fe [618] is 0.270-0.276 nm, which is considerably greater than the diameter of the iron atoms.

An electron-diffraction study of thin films of iron carbide has shown [619] that the phase transformation ε ⇌ ϰ takes place

TABLE 29. Composition and Properties of the ϰ-Phase of Iron Carbide Formed by Tempering and by Plastic Deformation

Method of obtaining the ϰ-phase	Value of index x	Density, g/cm^3	Curie point		Magnetization at temperature	
			°C	°K	0°	20° K
Tempering	1.91	6.73	265	538	875	800
Deformation . . .	2.23	6.96	265	538	985	900

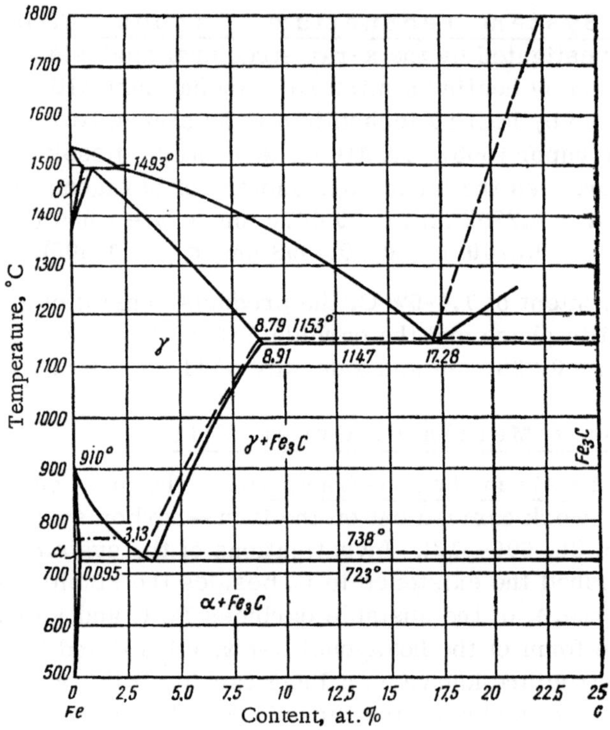

Fig. 60. Phase diagram of the system Fe–C.

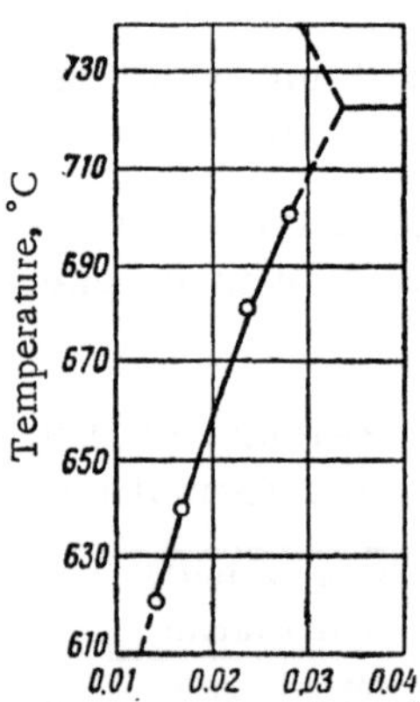

Fig. 61. Solubility of carbon in α-Fe.

at 380–400°C and the transformation ϰ → δ at 550°C. The temperature of the transition from the ε-carbide to the Hagg carbide is higher than 290°C, and the Hagg carbide changes into Fe_3C above 360°C [912]. The ε-carbide does not exist above 420°C.

There are some indications of the existence of the carbides FeC, FeC_2, and Fe_4C_3 [616], but these require confirmation.

The carbide Fe_xC (ϰ-phase) has been investigated by magnetic analysis [617]. From the magnetic diagrams the density, composition, and magnetization have been calculated for the carbide formed on tempering steel U11 at 300°C and by plastic deformation. The data calculated are given in Table 29.

The results obtained confirm that the composition of the $\varkappa$-carbide corresponds to the formula Fe_2C.

An electron-diffraction study of iron carbides obtained by treating iron films on NaCl crystals with CO, a mixture of CO and H_2, or acetylene [616] has revealed the existence of a cubic carbide Fe_4C with the carbon in a tetrahedron with four iron atoms displaced along the space diagonals from their ideal positions (Fig. 10). The constancy of the lattice parameter of Fe_4C, the tetrahedral coordination of the carbon atoms in the lattice, and the small distance Fe — C (for austenite it is 0.186 nm) lead to the conclusion that the carbide concerned is a chemical compound and not a solid solution.

A phase diagram of the system Fe — C based on numerous researches [621-636] is given in Fig. 60; in this the continuous lines indicate the metastable equilibrium diagram Fe — Fe_3C, while the broken lines indicate the stable equilibrium diagram between the melt and graphite. The presence of the metastable system, in addition to the stable one, is due [620] to the difference in rate of formation and growth of the cementite nuclei and the corresponding graphite nuclei.

The solubility of carbon in α-iron at room temperature is not more than 0.05% [637-639]. The solubility at high temperatures has been investigated on high-purity iron specimens, using the methods of x-ray analysis [640, 641], diffusion [642], measurement of the internal friction in solid solutions of carbon in α-iron [643], measurement of the relaxation of elastic stresses [644], and study of the equilibrium of the reaction $C + CO_2 \rightleftharpoons 2CO$ on passing CO over iron containing 0.014% C [645].

On the basis of this study of the equilibrium of the reaction, the following temperature dependence of the solubility of carbon in α-iron has been derived for temperatures up to 728°C [645]:

$$\log(\%\,C)_{\alpha}^{H} = -\frac{4509}{T} - 2.25 \cdot 10^{-4}\,T + 3.22.$$

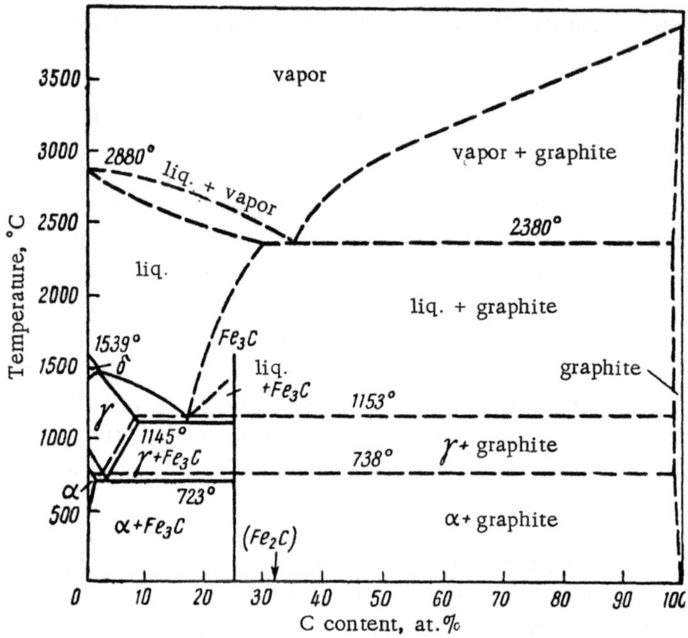

Fig. 62. Complete phase diagram of the system Fe — C.

Investigations using the method of tagged atoms (^{14}C) have shown [648] the temperature dependence of the solubility of carbon in α-iron which is given in Fig. 61. Extrapolation of the curve gives a solubility at 728°C of 0.035% C, in good agreement with the data in [645, 646].

It has been confirmed in [649, 650] that martensite is a solid solution of carbon in α-iron, while the reversibility of the martensite transformations in Fe — C alloys on heating has been demonstrated in [651, 652].

The mechanism and kinetics of the austenite-to-martensite transformation have not yet been finally established, and there are considerable differences between the various explanations of it [371, 653-655].

In studying the system Fe — C, a variant of the complete phase diagram up to 100% C, and including the vapor phase, has been proposed in [862] (Fig. 62); according to this diagram, there are no carbides richer in carbon than Fe_3C. The temperature of 2380°C is the boiling point of molten iron saturated with carbon.

The following structure of the carbide Fe_3C has been suggested [618]: the carbon atom forms a double bond with iron atoms I (the distance Fe — C is equal to 0.185 and 0.189 nm) and two bonds with iron atoms II (the distance Fe — C is equal to 0.206 and 0.215 nm). The distance $Fe_{II} - Fe_{II}$ is 0.249 nm. The carbide Fe_2C also crystallizes in a rhombic cell, the parameters of which bear a simple ratio to those of Fe_3C, namely,

$$a_{Fe_2C} \approx 2a_{Fe_3C},$$
$$b_{Fe_2C} \approx 3b_{Fe_3C},$$
$$c_{Fe_2C} \approx 1c_{Fe_3C}.$$

The reduction in carbon content in the Fe_2C cell as compared with Fe_3C does not cause any significant contraction of the lattice [371]. The carbide Fe_3C is a hard and brittle compound; at ordinary temperatures it is ferromagnetic, its Curie point lying in the range 210–215°C.

The specific electrical resistivity of iron — carbon alloys has been investigated in [665], and a relationship has been established between the logarithm of the specific resistivity and the Fe_3C content in the alloys. This is described by the equation

$$\log \rho = 0.90\mathrm{Fe_3C} + 0.996.$$

C content, %	0.06	0.17	0.53	0.83	1.12
Specific electrical resistivity, μΩ · cm . . .	10.45	10.57	11.87	12.94	13.90
C content, %	1.23	1.26	3.74	4.33	6.77
Specific electrical resistivity, μΩ · cm . . .	14.43	14.63	31.36	41.17	79.5

The carbide Fe_3C was first obtained by melting iron with sugar carbon or by heating a mixture of them at 700–800°C [64]. On melting a mixture containing some excess of carbon in a Kryptol furnace, the product obtained had 48% Fe_3C. Iron carbide has been produced at temperatures below the melting point by heating a mixture of pure reduced iron with sugar carbon in the proportions 5 : 1 in an electric resistance furnace at 920°C for 62 h in a hydrogen atmosphere. The carbide content in the sintered mass so formed is 13%.

In both cases the pure carbide is separated either by anodic solution in a 7% solution of hydrochloric acid or by treatment with dilute acids for three weeks. The solvents used were 2 *N* H_2SO_4, 7% HCl, and 60% CH_3COOH. The pure carbide was purified and precipitated by a 1 *N* solution of acetic acid over a period of four weeks [656].

The most widely used method for producing iron carbide is by the reaction of iron with carbon-containing gases. Thus, the carbide Fe_3C is obtained by the prolonged saturation of electrolytic sheet iron by benzene vapors at 700°C, the carbide being separated in pure form by anodic solution in a bath containing $FeCl_2$. The coarsely crystalline Fe_3C powder obtained is washed with dilute acetic acid, water, alcohol, and ether and then dried in vacuum [82].

In [618] pure Fe_3C was obtained by heating reduced iron or Fe_2O_3 in a mixture of H_2 and CO (15% CO) for 16 h at 500–550°C.

In order to investigate certain properties of the Hagg carbide, iron carbide was produced by treating high-purity iron or iron oxide with a mixture of CO and H_2. At ratios of $CO:H_2 = 1:4$, only the Hagg carbide is formed in the temperature range 300–360°C; a mixture of $Fe_{20}C_9$ and Fe_3C is formed at 380–460°C and pure Fe_3C at 460–560°C [947].

The most widely used method for obtaining pure iron carbides is by their chemical separation from steels and ferroalloys. This method can be used only for obtaining small quantities of the carbides for experimental purposes.

One group of methods is based on the difference in solubility between the carbide and the constituents of the metallic base.

The carbide Fe_3C has been separated by treatment with dilute hydrochloric, sulfuric, acetic, and formic acids and mixtures of them [58, 658, 659]. To prevent oxidation of the separated carbide, the separation and drying of the deposit are carried out with the exclusion of air. Acid separation was supplemented by magnetic removal of impurities. The most complete separation was obtained on treatment with a 0.1 *N* solution of sulfuric acid, but in this case the output did not exceed 35%.

In electrolytic separation the steel specimen acts as the anode. On electrolysis the metal dissolves, leaving the carbide as an insoluble residue [57, 58, 658, 659, 661, 662].

The electrolytes used consist of a 0.2 *N* solution of sodium sulfite with addition of 5% sodium citrate [58], a 1 *N* solution of potassium chloride with addition of 0.5% citric acid [661], a 15% solution of sodium citrate with addition of 1.2% KBr and 0.1% KI [659], and a 1 *N* solution of hydrochloric acid with addition of 2% sodium sulfite [57]. Electrolysis is carried out with cooling, the required current density being 0.02–0.03 A/cm^2.

The System Cobalt — Carbon. The first investigations of the cobalt — carbon system [666] showed that the maximum solubility of carbon in cobalt is 3.9% at 1700°C. A eutectic containing 2.9% C has been found at 1309°C [153]. At 2100°C the composition of the alloy corresponds to the formula Co_3C. This carbide exists in the melt and on cooling decomposes into cobalt and graphite. The authors also obtained alloys containing more carbon than corresponds to the formula Co_3C, from which the existence of higher cobalt carbides may be inferred.

Later investigations of the system Co — C have shown [667] that the high-temperature region in this system is similar to that in the system Ni — C. A eutectic has been found at 1300–1350°C; it is formed between the solid solution of carbon in cobalt and graphite and contains 2.8–3.0% C; there is also a carbide Co_3C which is somewhat more stable than Ni_3C. The assumed existence of a higher cobalt carbide has been confirmed in [657, 660, 668]. A carbide Co_2C, which is isostructural with Co_2N, has been obtained by the reaction of carbon monoxide with cobalt at 226–230°C [669].

The Co — C system has been investigated in the high-carbon, high-temperature region in [867]. It is shown there that a metastable system Co — Co_3C is superimposed on the stable cobalt–graphite system, as in the case of the Fe — C system. The instability is due to the fact that the carbide Co_3C readily decomposes on heating to 300–350°C (Fig. 63).

An investigation of the equilibrium of the system methane — carbon over cobalt used as a catalyst at 350–900°C has shown [657] that at 350–720°C, in addition to the reaction $CH_4 \rightleftharpoons C + 2H_2$, carbide is formed in accordance with the equation

$$n\mathrm{Co} + \mathrm{CH_4} \rightleftarrows \mathrm{Co}_n\mathrm{C} + 2\mathrm{H_2} - 47.26\ \mathrm{kJ}\ (11.28\ \mathrm{kcal}).$$

At 428–900°C cobalt is saturated with carbon at low methane concentrations with the formation of the solid solution Co–Co_nC. The equilibrium methane content increases with rise in the carbon content in the carbide, attaining its maximum when the pure carbide phase is reached.

The System Nickel – Carbon. From the first investigations of the system Ni – C [85, 577, 667, 671, 672] it was possible to construct a phase diagram based on the graphical interpolation of thermal-analysis results and the determination of the solubility of carbon in nickel in the liquid state. According to this diagram, a eutectic containing 2.2 wt.% (0.9 at.%) C is formed at 1314°C.

The solubility of carbon in nickel in the solid state is 0.55 wt.% (2.7 at.%) C at 1318°C. Below are given data on the solubility of carbon in nickel in the solid state [674]:

Temperature, °C	700	745	925	1055	1170	1265
Solubility:						
wt.%	0.08	0.1	0.2	0.3	0.4	0.5
at.%	0.4	0.48	0.95	1.45	1.95	2.4

The solubility of carbon in nickel in the liquid state at 2100°C is 6.4%, which corresponds to the carbide Ni_3C (6.39% C). On this evidence it is concluded that the carbide Ni_3C exists. At a temperature of 2490°C and a pressure of 3999 N/m^2 (30 mm Hg), liquid Ni_3C boils and decomposes with separation of graphite and metal vapor.

In an attempt to obtain nickel carbide with a higher carbon content [676], it was found, by means of x rays, that only the lines due to Ni_3C were observed in all the products.

Ni – C alloys have been thoroughly investigated in [677, 678]. The phase diagram has been constructed on the basis of metallographic analysis, electrical-resistance measurements in the solid and liquid states, and thermal analysis. From the result it has been possible to determine more accurately the position of the liquidus lines on the diagram. The eutectic point, for which the re-

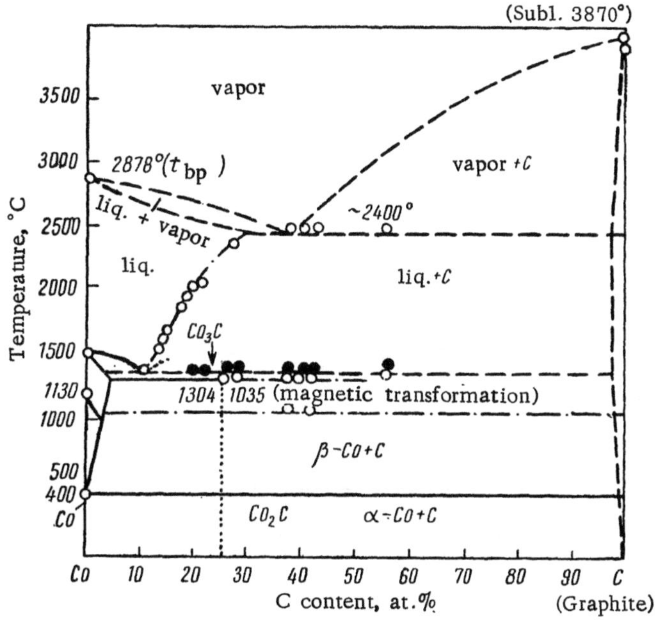

Fig. 63. Phase diagram of the system Co – C [867].

sults of thermal analysis and a discontinuity in the resistance measurements gave the same position, corresponds to 1.3% C, which is at variance with the data given above from [577, 672]. This discrepancy is explained by the results given in [679], in which it was shown that at high rates of cooling Ni – C alloys may crystallize in accordance with the metastable phase diagram, with the separation of Ni_3C. In these circumstances the eutectic point is displaced in the higher carbon direction. Thus only one carbide Ni_3C is found in the system Ni – C, and this is stable at temperatures above 1600°C and below 300°C [371].

The Ni – C system has been investigated in [867] in the composition range up to 69.2 at.% C and temperatures up to 3000°C. As in the system Fe – C and Co – C, there is, in addition to the stable phase diagram, a metastable Ni – Ni_3C system owing to the low decomposition temperature of the carbide (200–400°C) (Fig. 64).

Nickel carbide is a gray or black powder which completely decomposes at 370°C.

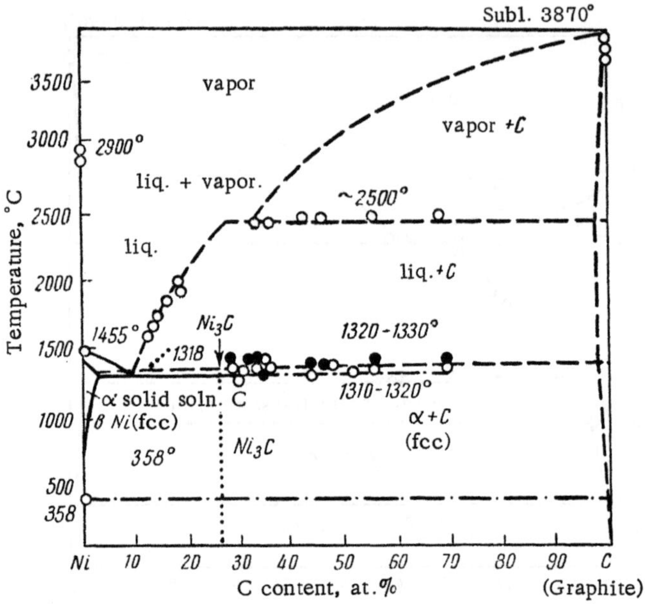

Fig. 64. Phase diagram of the system Ni – C [867].

The carbide Ni_3C has been obtained by treating nickel oxide with carbon [676, 680, 681]. The oxide is first reduced with hydrogen at 320°C, and then carbon monoxide is passed over the reduced nickel powder for 170 h at 240–250°C. The process is controlled by weighing the carbon dioxide formed, which is adsorbed by caustic soda. Methane or acetylene is sometimes used instead of CO.

Investigations into the reduction of nickel oxide with solid carbon have shown [675] that the rate of reduction, which is insignificant at 800°C, rises sharply at 900°C.

The System Ruthenium–Carbon. Liquid ruthenium dissolves carbon, which on cooling is precipitated in the form of graphite [66, 85]. The solubility of carbon in molten ruthenium increases with rise in temperature, and at the boiling point of the metal it is 4.8%.

It has been shown from an x-ray study of the products of the reaction of ruthenium with carbon [771, 772] that on heating a mixture of ruthenium and carbon black in the proportions of 1 : 10 in a

TABLE 30. Phase Compositions of Reaction Products of Osmium and Carbon Black [819]

Molar ratio Os:C	Temp., °C	Time, min	Phase composition from x-ray data
1:10	2200	240	Os + graphite
1:10	2400	240	Os + graphite
1:10	2600	180	Os + graphite
Ta:Os:C 1:10:100	2600	90	Os + graphite + TaC

helium atmosphere at 2600°C for 4 h, products are formed on whose x-ray diffraction diagrams there are lines of a new phase isostructural with WC and having lattice parameters $a = 0.290785$ and $c = 0.282186$ nm and with a density of 9.008 g/cm³. This phase has been identified as RuC.

An attempt was subsequently made to confirm the existence of ruthenium carbide [819]. On heating a mixture of ruthenium and carbon black in the proportions of 1:10 at 2600°C for 2 h, the authors obtained products whose x-ray diffraction diagrams revealed only lines from ruthenium and graphite.

The System Rhodium—Carbon. The solubility of carbon in molten rhodium is 1.42-7.28%. On cooling, the carbon is precipitated in the form of graphite [67, 85].

The System Palladium—Carbon. Molten palladium on heating dissolves carbon, which on subsequent cooling is precipitated in the form of graphite. The solubility of carbon is 2.45% [66, 85].

There are data to show that methane or a mixture of methane and hydrogen reacts vigorously with palladium powder to yield a product containing 4.4% C and corresponding to the formula Pd_5C_2 [773]. However, these data have not been confirmed.

The System Osmium—Carbon. At its boiling point osmium dissolves 3.9-4% C [85].

No carbides were found to form in an investigation of the products of the reaction of osmium powder with methane at 1200 and 2000°C [601].

An x-ray study of the reaction products of osmium and carbon black in a helium atmosphere at 2200-3000°C has shown that on heating a mixture of osmium and carbon in the proportions 1:10 at 2600°C for 4 h, a carbide OsC is formed: it has a hexagonal lattice with parameters a = 0.290769, c = 0.282182 nm, and a density of 16.25 g/cm^3 [771, 772]. However, these data were not confirmed in [819]. The phase compositions of the reaction products of osmium and carbon black are given in Table 30.

The System Iridium—Carbon. At its boiling point molten iridium dissolves 2.08% carbon [66, 85]. On cooling, the carbon is precipitated in the form of graphite.

The System Platinum—Carbon. Molten platinum dissolves 1.2-1.45% C, which is precipitated on cooling [66, 85]. On the alternate deposition of platinum and carbon on a thin colloidal film and on heating the product so obtained in vacuum at 1100°C, a film is formed, 10 nm thick, which in the authors' opinion is a carbide of platinum [85].

Chapter VI

Carbides of Elements of the Boron Group

The System Boron—Carbon. The first investigations of the system B — C [682] established the existence of the phase B_4C ($B_{12}C_3$). Numerous subsequent investigations of this system [683-690] have led to the final form of the phase diagram shown in Fig. 65. In particular, the diagram confirms the existence of the carbide $B_{12}C_3$ with a large homogeneity range (17.6-29.5 wt.% C), the carbide $B_{13}C_2$, and a new phase.

A metallographic study of the specimens has revealed the existence in the range 6.4-10% C of a chemical compound containing about 8% C and corresponding in composition to $B_{12}C$. X-ray diffraction diagrams of specimens similar in composition to $B_{12}C$ have an individual character indicating the complex structure of this phase.

In an investigation of the diffusion of boron into carbon [696], it was found that the diffusion rate is greater than that for carbon into boron. The temperature dependence of the coefficient of diffusion of boron into carbon is represented by the equation

$$D = 3.02\, e^{\frac{-28625}{T}}$$

The carbide B_4C is a dark gray powder which crystallizes in the rhombohedral lattice. It is characterized by great hardness, surpassed only by that of diamond [48,000 MN/m^2 (4950 kgf/mm^2)]. The abrasive power of the powder, determined in arbitrary units, is 0.613 for diamond, 0.442 for boron carbide, and 0.314 for silicon

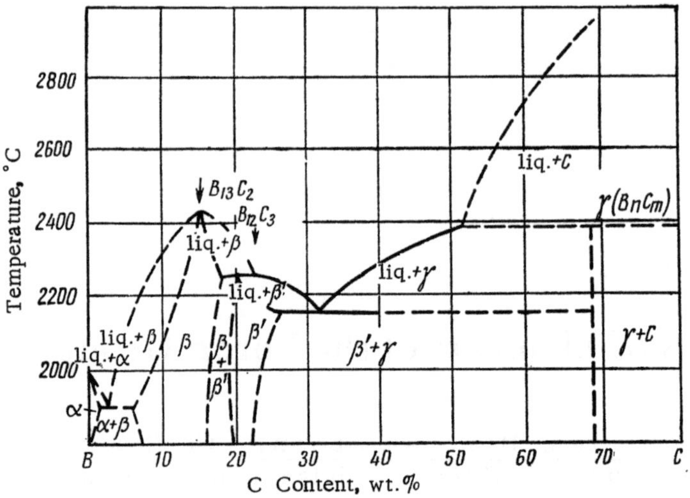

Fig. 65. Phase diagram of the system B — C [690].

carbide [694, 695]. The abrasive power falls sharply with increase in free carbon content in the boron carbide.

The carbide B_4C_{1-x} is characterized by a linear chain —C—C—C— and is a hole semiconductor. The carbide $B_{13}C_2$ is characterized by a linear chain —C—B—C— and exhibits electron conduction.

Measurements of the specific electrical resistivity and thermo-emf of boron carbide in relation to temperature show that up to 500°C the resistivity falls almost linearly, while at higher temperatures it decreases more slowly. The thermo-emf varies inversely. The width of the forbidden band of carbide specimens containing 78.7% B and 21.7% C is $0.16–0.208 \cdot 10^{-19}$ J (0.1–0.13 eV). The electrical resistance of commercial boron carbide is lower than that of the pure carbide by almost an order of magnitude.

In [379], a study was made of the evaporation of boron carbide and pure boron in the temperature range 1750–2500°C, and the standard heat of dissociation of boron carbide into gaseous boron and graphite was determined as 553 kJ (132 kcal). This is in good agreement with the data in [376].

The production of boron carbide by synthesis from its elements is carried out by heating a mixture of boron and carbon black at temperatures above 1600°C. By this method it is possible to obtain a fairly pure product, but in economic terms it is of low efficiency and has no practical importance.

The most widely used method for producing boron carbide is by the reduction of boric anhydride with carbon black in accordance with the reaction

$$2B_2O_3 + 7C = B_4C + 6CO.$$

This process takes place in two stages:

$$B_2O_3 + 3CO = 2B + 3CO_2,$$
$$4B + C = B_4C.$$

In the reaction of B_2O_3 with C a layer of boron carbide (through which diffusion takes place) is formed from the molten mixture on pieces of graphite [482]; boron is formed by the reduction of B_2O_3 with carbon until the graphite is completely transformed into carbide.

The temperature at which the reduction of boric anhydride by carbon monoxide begins is above 1320°C, the rate of reduction increasing with rise in temperature. However, it must be borne in mind that boron carbide decomposes peritectically at 2250°C. The boron volatilizes and the remaining product is enriched in carbon.

A method for producing boron carbide has been developed in which a charge consisting of boric acid and carbon black in the form of briquets is dehydrated by heating at 800°C; in this process the boric anhydride is melted and a sponge is formed consisting of boric anhydride well mixed with carbon. Subsequently the sponge is heated, the commercial methods of producing boron carbide differing mainly in the type of furnace in which the process is carried out. These may be cored or coreless, Tammann-type furnaces, or electric arc furnaces. In the cored and coreless furnaces the temperature is not uniformly distributed throughout the three zones into which the charge can be divided: (1) the center, high-temperature zone, where the temperature exceeds 1850°C (in this zone the reduction of boric anhydride with the formation of carbide is completed); (2) an intermediate zone containing a mixture of unreduced boric anhydride, boron carbide, and unused carbon; (3) an outer zone consisting mainly of the initial charge. The products of the second and third zones are used as raw material for the subsequent charge, while the product of the reaction occurs mainly in the first zone.

The magnesiothermic method of obtaining boron carbide [692, 693] is based on the reaction

$$2B_2O_3 + 6Mg + C = B_4C + 6MgO.$$

It is carried out by heating suitable charges in an electric resistance furnace with graphite heaters in a hydrogen atmosphere at 1800°C. As the reaction products contain magnesium oxide, they are washed with hydrochloric acid and the residue is heated in vacuum at 1800°C to remove volatile impurities. This method has not found wide industrial application, but it can be used to obtain carbide of the exact stoichiometric composition.

Boron chloride is reduced by hydrogen in the presence of carbon according to the reaction

$$4BCl_3 + 6H_2 + C = B_4C + 12HCl$$

on heating at 600°C a charge consisting of boric anhydride and excess carbon in a stream of chlorine. In this way a mixture of BCl_3 and CO is formed and this gaseous mixture, together with hydrogen, is passed over a heated tungsten filament on which a layer of the carbide is deposited.

In [863] the conditions for obtaining boron carbide single crystals by deposition from the gaseous phase in accordance with the reaction

$$4BCl_3 + CCl_4 + 8H_2 = B_4C + 16HCl$$

have been investigated, using the apparatus shown in Fig. 66. Single crystals, 3–4 mm long, in the form of needles and platelets were obtained by deposition at 1550–1650°C for 4–5 h, the rate of supply of the mixture of BCl_3 and CCl_4 being 1 g/min with excess of hydrogen. The impurity content in the products obtained was $5 \cdot 10^{-4}$%, the impurities being mainly silicon and magnesium.

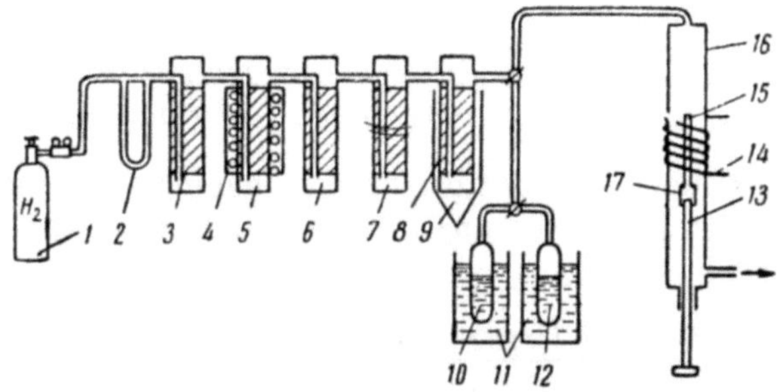

Fig. 66. Apparatus for the deposition of boron carbide from the gaseous phase. 1) Hydrogen cylinder; 2) flowmeter; 3) flask containing KOH; 4) furnace; 5) palladinized asbestos; 6) flask containing KOH; 7) flask containing P_2O_5; 8) activated carbon; 9) vessel containing liquid nitrogen; 10, 12) containers with BCl_3 and CCl_4; 11) thermostat; 13) copper rod; 14) coil; 15) graphite rod; 16) quartz tube; 17) ceramic connector.

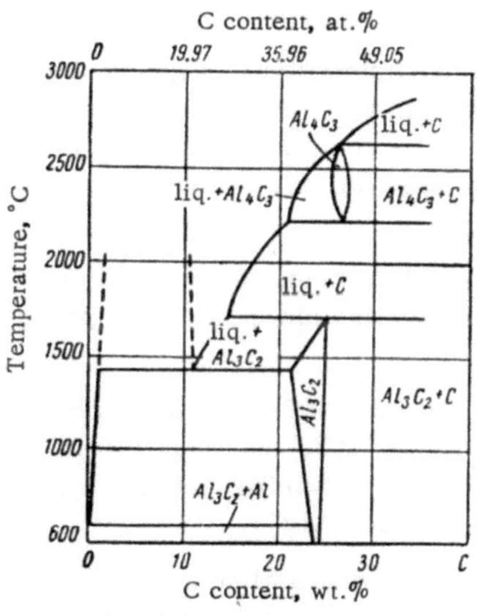

Fig. 67. Phase diagram of the system Al – C [86].

Dense specimens of boron carbide are obtained chiefly by hot compaction at temperatures above 2000°C and a pressure of 3390 MN/m^2 (350 kgf/cm^2) [692]. A porosity of the specimens approaching zero is attained with these powders at temperatures of 1650–1700°C and a pressure of 1940 MN/m^2 (200 kgf/cm^2).

The chemical composition changes during hot compaction, the impurity content decreasing and the relative content of boron and carbon increasing.

The System Aluminum—Carbon. The phase diagram of this system has not yet been completely determined, and the part of the diagram shown in Fig. 67 is only hypothetical [86]. According to this diagram, the system Al—C contains two carbides: Al_3C_2 and Al_4C_3. The existence of a carbide Al_3C has been suggested in [85]. However, numerous investigations have shown that the only stable carbide in the system is Al_4C_3.

The solubility of carbon in aluminum is 0.02–0.04% at 1300–1500°C; at 1000–1100°C it is practically zero [161].

The carbide Al_4C_3 is a yellow powder [86]. In vacuum it is stable up to 1200°C, but at 2200°C it sublimes without melting.

A mass-spectrographic study of gaseous aluminum carbide in equilibrium with aluminum vapor and graphite has shown [170] that at 1800°C the vapors contain aluminum atoms and molecules of Al_2 and Al_2C_2, the peak heights of the ion masses being in the ratio 1050 : 0.1 : 1.

The tendency of aluminum carbide to dissociate in vacuum is utilized in the semicontinuous method for producing pure aluminum.

The values of the heat of formation of aluminum carbide obtained by various authors [103, 169, 171, 175] differ considerably owing to the difficulty of obtaining carbide which exactly corresponds to the composition Al_4C_3.

In [163] an amorphous aluminum carbide is described, whose formation from its components took place on the cold walls of an arc furnace. This carbide is thermally less stable than that usually obtained by synthesis. It decomposes at 1200°C and is characterized by lower chemical stability. Its heat of formation has been shown to be practically identical with that of the crystalline carbide.

TABLE 31. X-Ray Analysis Data

Phase	Lattice parameters, nm		c/a	Density, g/cm^3
	a	c		
Ti_2InC	0.313_2	1.406	4.490	6.19
Ti_2TlC	0.315_8	1.398	4.426	8.59
Nb_2InC	0.317_2	1.437	4.53_0	8.29
Ta_2GaC	0.310_4	1.357	4.37_1	13.01

Evidently the greater reaction activity of this carbide is due solely to its greater dispersion.

The carbothermic reduction of aluminum oxide in the temperature range 1400–1900°C has been investigated in [163], and it has been found that the oxide is reduced by carbon to the carbide Al_4C_3 via intermediate stages in which the oxycarbides Al_4O_4C and Al_2OC are formed.

Aluminum carbide has also been obtained by heating aluminum with calcium carbide, by reducing aluminum oxide with calcium carbide in an electric arc furnace, by reducing sodium aluminate with carbon [64], by heating a mixture of aluminum and carbon up to 2000°C [164], and by heating aluminum melted in a carbon crucible to 1700°C [161, 165]. Aluminum carbide is also formed during the electrolysis of cryolite-alumina melts [167].

According to [166], in this process an important part is played by the nature of the carbonaceous materials contained in the hearth blocks and the anode mass. The authors have investigated the effect of various electrode materials on the products of their reaction with aluminum powder in vacuum at 1000°C. Below are given the results of investigations that demonstrate the effect of the source of the graphite on its tendency to carbide formation:

Gray natural graphite	43.2	%	Al_4C_3
Black natural graphite	33.6	"	"
Artificial graphite	44.7	"	"
Anthracite	36.2	"	"
Pyrolytic petroleum coke	35.3	"	"
Cracking petroleum coke	33.18	"	"
Pitch coke	44.4	"	"
Foundry coke	25.4	"	"

Artificial graphite has the greatest tendency to carbide formation and petroleum and foundry cokes the least tendency.

Gallium, Indium, and Thallium do not form compounds with carbon. However, the ternary compounds Ti_3InC and Ti_3TlC are known [176]; they were obtained by the reaction of titanium carbide with metallic indium and thallium. They crystallize in a cubic lattice with parameters a = 0.4199 nm for Ti_3InC and a = 0.4209 nm for Ti_3TlC. The phases Zr_2TlC, Hf_2TlC [177], Ti_2TlC, Nb_2InC, and Ta_2GaC [176] have also been obtained. These phases were produced by heating mixtures of the carbides with the metals at 850°C for 600–900 h.

The results of an x-ray study of the phases obtained are given in Table 31.

Carbide phases containing rare-earth metals of the type La_3InC, Pr_3GaC, Nd_3GaC, Nd_3InC, and Nd_3TlC have also been synthesized [783].

Chapter VII

Silicon Carbide

The existence of only one compound, SiC, has been established in the system Si — C. Reports in the literature [698, 699] with regard to the existence of the carbides Si_2C and SiC_2 have not been confirmed.

The investigation of the phase diagram of the system Si — C is very difficult owing to the high chemical activity and vapor tension of molten silicon and to the existence of various crystalline modifications of SiC. In [700] two possible versions of the phase diagram are given; these do not take account of the crystalline modifications of SiC (Fig. 68).

The maximum solubility of carbon in silicon is 0.6 wt.% (1.4 at.%) [701].

The investigation of alloys in the composition range Si — SiC has revealed the existence of two phases: a solid solution of carbon in silicon and silicon carbide. In specimens whose composition lies within the range SiC — C the presence of SiC and carbon is observed; prolonged heating in vacuum at 2000-2100°C leads to the dissociation of silicon carbide, to the evaporation of silicon, and to the separation of free carbon.

Thus the investigations in [701, 702] confirm the accuracy of the diagram shown in Fig. 68a.

The existence of only one compound SiC in the system Si — C has been confirmed in [702]; it is formed peritectically at 2830 ± 40°C. The solubility of carbon in liquid silicon, according to the same source, is not more than $5 \cdot 10^{-3}$ at.%.

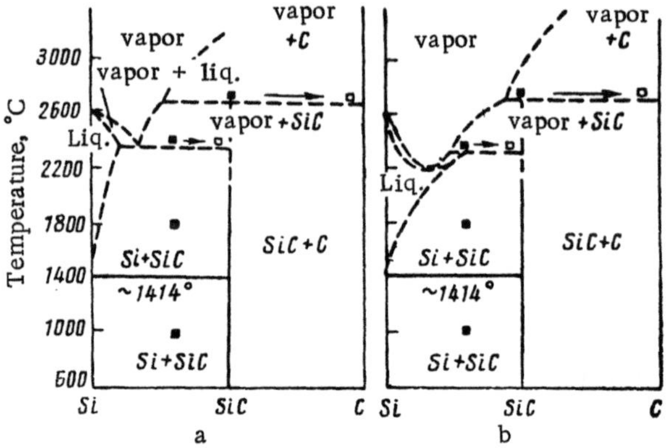

Fig. 68. Phase diagrams of the system Si – C.

Because silicon carbide has been dealt with in many papers [694, 702, 704-711, 835], we shall give briefly only the principal data about the properties of silicon carbide and the methods for producing it.

Silicon carbide exists in the form of the cubic β-SiC and the hexagonal α-SiC. The transition temperature from one modification to the other has not been determined. Some workers consider that β-SiC is unstable above 2000°C and is converted into α-SiC [726, 727]. According to [728], α-SiC may be formed below 2100°C side by side with β-SiC; in [728] both modifications were obtained at 2700°C.

However, it was found in [729] that β-SiC is usually formed first, the transition of the β-modification into α taking place at high temperatures through the gaseous phase. This was confirmed in [730], in which the rate constants of the transformation β-SiC → α-SiC at 2100 and 2300°C were calculated, as well as the activation energy of the transformation, which is 662 kJ/mole (158 kcal/mole). This is similar to the heat of evaporation of SiC in the same temperature range, namely 557 kJ/mole (133 kcal/mole).

It has been shown experimentally in [943] that a certain excess pressure of the inert gas (~20 atm of argon) leads mainly to the formation of β-SiC even at 2850°C, i.e., the transformation of the β- to the α-modification is suppressed at high temperatures.

Nitrogen, by penetrating into the carbon sublattice, suppresses the $\beta \rightarrow \alpha$ transformation and stabilizes the β-SiC phase [944].

The transition of the α- to the β-modification is not usually observed. β-SiC possesses one crystalline form, having a unit cell of the diamond type; it has great hardness and wear resistance [323, 731, 944].

α-SiC forms many structural types with hexagonal or rhombohedral lattices [816-818]. The layers of silicon and carbon atoms in them are displaced in different ways relative to one another and form hexagonal packing from the tetrahedral structures in which atoms of each occupy half the vacant sites in the other. The individual tetrahedra consisting of layers of carbon and silicon atoms are situated parallel and antiparallel, forming stacks of layers in the same direction [732].

The principal structures of hexagonal SiC are regarded as SiC-II and SiC-III [733, 734], all the remaining modifications consisting of various combinations of these types. Numerical symbols have been suggested for these, the sequence of numbers indicating the number of elementary layers with parallel and antiparallel orientations of the tetrahedra situated in the stacks.

The origin of the various forms of SiC is connected with the formation of defects on the surface of the SiC crystals during their growth. The reason for the various forms is said in [735] to lie in the difference in size of the carbide particles in the vapor state. However, these data are not confirmed by the actual mechanism of formation of SiC.

The effect of impurities adsorbed by the growing surface of the crystals is a more likely cause. Thus it was found in [683] that SiC-III is formed in the presence of boron as an impurity, whereas SiC-II is formed in its absence. Chemically pure silicon carbide is colorless, but the commercial carbide may be green or black in color. The green carbide consists mainly of SiC-II and the black form consists of a mixture of 60% SiC-II and 40% SiC-III.

The dissociation of silicon carbide takes place at 2700°C [736, 737]. An x-ray study of the carbide heated in vacuum has shown [738] that at 2050°C α-SiC partially decomposes with precipitation of graphite. At 2150°C complete decomposition occurs.

In [702, 740] the partial pressures in the systems SiC — C and Si — SiC were determined by means of the mass-spectrograph in an investigation of the equilibrium of the system Si — C by Knudsen's method, and the molecular composition of the gaseous products was determined. These products contained, in addition to silicon atoms and molecules of Si_2, Si_3, and Si_4, molecules of the gaseous carbides SiC, SiC_2, Si_2C, Si_2C_3, and Si_3C.

The values of the heat of formation, entropy, and heat of combustion of the various modifications are very similar [702, 741].

Numerous investigations of the electrophysical properties of silicon carbide, a review of which is given in [711, 712], have shown that the carbide is a semiconductor of the impurity type.

An investigation of the effect on silicon carbide specimens of irradiation by neutrons whose total flow was $7 \cdot 10^{18}$–$1 \cdot 10^{21}$ of fast neutrons/cm^3 [913] revealed an expansion of the specimens to a maximum extent of 2.9%. The lattice parameter a increased by 0.89% and c by 1.09%.

The most widely used and thoroughly investigated method of producing silicon carbide is by the reaction of silica with carbon. The overall reaction is represented by the equation

$$SiO_2 + 3C = SiC + 2CO.$$

It takes place in two stages: reduction of silica by carbon to give metallic silicon and the reaction of silicon vapor with carbon or carbon monoxide:

$$SiO + 2C = Si + 2CO,$$

$$Si + C = SiC,$$

or

$$3Si + 2CO = 2SiC + SiO_2.$$

The suggestion has been made [694, 704, 713] that an intermediate product may be silicon monoxide, which at the temperatures of the production facility is formed by the reaction

$$SiO_2 + C = SiO + CO.$$

TABLE 32. Phase Composition of Silicon Carbide in Relation to the Furnace Zones [714]

SiC	Place from which sample was taken	SiC	SiC—I	SiC—II	SiC—III
Green	Within core	34	0	66	0
	Intermediate zone	29	3	67	0
	Zone near surface	37	6	52	5
	Surface layer	80	Remainder		
Black	Within core	28	0	72	0
	Intermediate zone	22	1	72	5
	Zone near surface	8	5	18	69
	Surface layer	23	11	24	42

TABLE 33. Composition of Two Varieties of Silicon Carbide

Variety	Content, %						
	SiC	Si_{free}	C_{free}	Fe	Al	CaO	SiO_2
Green	98.8	0.59	0.03	0.11	0.06	0.01	—
Black	96.2	—	0.13	1.05	—	—	0.94

The raw material for producing silicon carbide consists of quartz sand containing not less than 98.5% SiO_2 and the minimum amount of impurities in the form of Al_2O_3, CaO, MgO, and Fe_2O_3. This is because the impurities alumina and magnesium and calcium oxides may form silicates which promote sintering of the charge, while iron oxide leads to the formation of iron—silicon alloys. The carbon-containing raw materials are low-ash coke, anthracite, and petroleum coke. Wood sawdust is added to the charge (to increase its gas permeability) together with sodium chloride, the effect of which is to refine the charge by chlorinating the impurities.

Electric resistance furnaces are used in the production of SiC. In the most advanced furnace available at present, the length is divided into sections insulated from one another by refractory brick. Electrodes are fixed to the end walls of the furnace and a carbon core is placed between them in the center of the furnace.

The phase composition of the silicon carbide obtained from the various furnace zones is given in Table 32.

By the method described it is possible to obtain commercial-quality silicon carbide whose composition is given in Table 33.

The high semiconducting properties of silicon carbide single crystals make it necessary to develop methods for producing these single crystals on a commercial scale. The following possible methods for obtaining the single crystals are given in [702]: (1) crystallization from liquid silicon carbide; (2) crystallization from solutions; (3) sublimation; (4) thermal decomposition; (5) thermal reduction.

The first method necessitates the development of a plant in which it is possible to attain a temperature of 2000°C at a pressure of hundreds of thousands of atmospheres, since, under ordinary conditions, silicon carbide cannot be converted into the liquid state.

The second method — crystallization from solutions — is fairly widely used. Crystals of n- and p-type SiC have been obtained from molten silicon into which additions of Li, Mg, B, Al, Ga, Ge, As, Sb, Cu, Fe, Ni, and Au have been introduced in various proportions at 1600°C. A study of the effect of these additions on the crystallization of SiC has shown that they stabilize α-SiC [702].

Crystals of SiC have been obtained [717] from silicon — nickel melts of eutectic composition at 1000°C and from silicon — aluminum — zinc melts at 525°C. SiC crystals have also been obtained from iron — silicon melts [718], silicon — carbon melts [719], and ferro — silicon — chromium [720].

In an investigation of the crystallization of SiC on melting silicon in graphite crucibles in a pure argon atmosphere, using high-frequency heating [781], it was possible to obtain crystals of cubic structure with a size less than 1 mm^2 at temperatures above 2000°C. If impure argon was used, crystals of hexagonal SiC were obtained.

The most widely used method of growing silicon carbide single crystals is by sublimation. In [711], a cylinder of SiC crystals was sintered and then heated in a graphite tube furnace. By

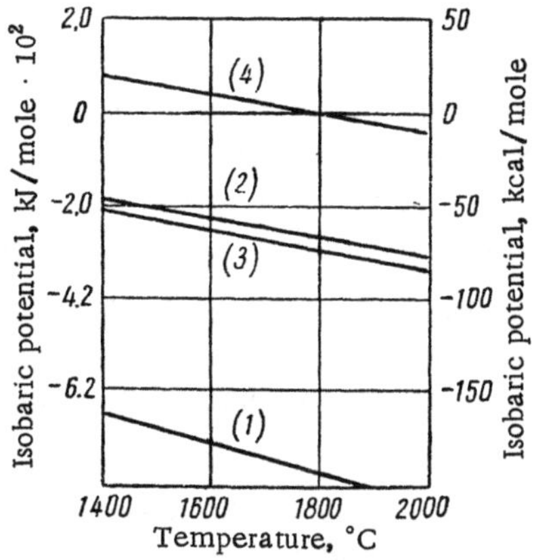

Fig. 69. Temperature dependence of the change in the isobaric potential for the reaction by which SiC is formed from the gaseous phase. The figures on the curves refer to the reaction number.

maintaining the cylinder in an inert atmosphere for several hours at 2500°C, thin platelets consisting of SiC single crystals were grown in the cylinder cavity. Purer crystals were grown by using silicon carbide obtained from a mixture of silicon and high-purity graphite [702]. In all these cases, however, the single crystals contained as an impurity nitrogen which had penetrated from the furnace through the wall of the graphite reaction vessel. The deposition of pyrographite on the walls of the graphite reaction vessel by the pyrolysis of methane built up a dense impermeable layer up to 50 μm thick.

The growth of crystals by sublimation of SiC in an argon atmosphere has been studied in [131]. The initial SiC powder was placed in the space between the graphite crucible and the graphite cylinder. The silicon carbide vapors entered the zone of crystal growth through holes in the cylinder walls, the deposition of the crystals being controlled by varying the diameter of the holes and the temperature.

Another method that is extensively used for producing silicon carbide crystals is by deposition from the gaseous phase by thermal decomposition of silicon- and carbon-containing compounds or by reduction. Thus, silicon carbide has been obtained in [53] from a mixture of silicon tetrachloride, toluene, and hydrogen.

In studying the conditions for growing single crystals by deposition from the gaseous phase, it has been shown [724] that β-SiC crystals are deposited at 2000°C on a carbon filament from a mixture of $SiCl_4$, C_7H_8, and H_2 with the first two in the proportion 20 :1. In the absence of hydrogen it is impossible to obtain SiC whatever the proportions present in the gaseous mixture.

SiC single crystals were obtained in [780] by passing a current through a carbon filament 0.175 mm in diameter. In this case the filament temperature reached 2300-2400°C, but it burned out after 15 min.

A thermodynamic analysis of the process involved in the formation of silicon carbide and of the gaseous phase has been made in [725]. The change in the free energy of the following reactions was calculated:

$$(SiCl_4)_{gas} + (CCl_4)_{gas} + 4H_{2(gas)} = SiC + (8HCl)_{gas},$$
$$(SiCl_4)_{gas} + {}^1/_6\,(C_6H_6)_{gas} + {}^3/_2\,(H_2)_{gas} = SiC + 4\,(HCl)_{gas},$$
$$(SiCl_4)_{gas} + {}^1/_7\,(C_7H_8)_{gas} + {}^{10}/_7\,(H_2)_{gas} = SiC + 4\,(HCl)_{gas},$$
$$(SiCl_4)_{gas} + {}^1/_7\,(C_7H_8)_{gas} = SiC + {}^8/_7\,(HCl)_{gas} + {}^{10}/_7\,(Cl_2)_{gas}.$$

The results given in Fig. 69 confirm that SiC cannot be formed in the absence of hydrogen.

The changes in free energy of the reactions by which carbides of possible impurities may be formed have also been calculated:

$$21FeCl_3 + C_7H_8 + 27.5H_2 = 7Fe_3C + 63HCl,$$
$$28AlCl_3 + 3C_7H_8 + 15H_2 = 7Al_4C_3 + 54HCl,$$
$$7TiCl_4 + C_7H_8 + 10H_2 = 7TiC + 28HCl,$$
$$28BCl_3 + C_7H_8 + 38H_2 = 7B_4C + 84HCl.$$

The calculated data showed that in the temperature range in which silicon carbide is formed the reactions (5)-(8) also take place. As a result, to obtain pure SiC it is essential to eliminate carefully all impurities from the raw materials. Meerson et al. [725] have investigated the effect of the temperature of the deposition surface, the concentration of the reagents, and the rate of gas flow on the rate of deposition of silicon carbide. From this work the optimum conditions for obtaining silicon carbide from a mixture of silicon tetrachloride, toluene, and hydrogen have been established, namely: temperature, 1600-1700°C; ratio of $SiCl_4 : C$ in the initial gaseous mixture, 3 : 1; 8-11 times as much hydrogen as the calculated amount; rate of gas flow of 1.0-1.5 liters/min for a reactor 50 mm in diameter.

Chapter VIII

Chemical Properties of Carbides

Lithium Carbide (Li_2C_2) easily ignites at room temperature in a fluorine or chlorine atmosphere with the formation of lithium fluoride or chloride. With bromine and iodine reaction takes place on heating [64]. In an atmosphere of oxygen, sulfur, or selenium vapor it burns at 700-800°C; with phosphorus it forms the phosphide, which is decomposed by water with evolution of phosphine.

Lithium carbide is decomposed by water with evolution of acetylene. The reaction proceeds rapidly at room temperature and very vigorously on heating to 100°C. It has been shown in [4] that the pure carbide is completely decomposed in water with the evolution of an amount of acetylene exactly corresponding to the composition Li_2C_2. Concentrated acids act slowly on the carbide; molten alkali decomposes it with the evolution of heat, while it is rapidly decomposed by molten potassium chlorate, nitrate, and permanganate.

The carbide Li_2C_2 is a powerful reducing agent; it reduces LiCl to Li_2Cl with the separation of carbon. The decomposition $Li_2Cl \rightarrow LiCl + Li$ occurs at high temperatures. In this case the metal then reacts with carbon to form Li_2C_2. In the analysis of lithium carbide [4], the carbon content is determined by measuring the volume of acetylene evolved during its decomposition by water, in a graduated tube over a layer of mercury. The lithium content is determined in the solution by a volumetric method.

Potassium Carbide (K_2C_2) is decomposed by water vapor with the evolution of acetylene. In a chlorine atmosphere it burns at room temperature; it does not react with CO at ordinary temperatures.

KHC_2 reacts vigorously with CO_2, SO_2, and chlorine [64].

Sodium Carbide (Na_2C_2) is not affected by dry air, but in moist air it is decomposed with the evolution of acetylene:

$$Na_2C_2 + 2H_2O = 2NaOH + C_2H_2.$$

In the presence of water, decomposition takes place explosively with the separation of carbon [64].

Chlorine and bromine also react explosively with Na_2C_2 with the formation of sodium chloride or bromide and the separation of carbon. The carbide reacts less vigorously with iodine to form C_2I_4. On heating the carbide with phosphorus in a hydrogen atmosphere, sodium phosphide is formed, which in the presence of water liberates phosphine. In a stream of hydrogen chloride the carbide burns with the formation of sodium chloride and the separation of carbon and hydrogen.

Hydrogen sulfide reacts with Na_2C_2 at 150°C; in an atmosphere of CO_2 or SO_3, sodium carbide ignites even at room temperature; CO has no effect up to 250°C.

Such oxidizing agents as chlorates, nitrates, and bromates form with the carbide mixtures which ignite under impact or on abrasion; the chlorides and iodides of mercury and lead and the chlorides of aluminum and iron decompose explosively on grinding with the carbide.

Alcohols decompose the carbide with the evolution of acetylene and the formation of the corresponding alcoholates of sodium.

The compound $NaHC_2$ is readily decomposed by water [5]: $NaHC_2 + H_2O = C_2H_2 + NaOH + 60.7$ J. $NaHC_2$ is decomposed at room temperature in an atmosphere of chlorine or bromine; it is readily decomposed by alcohol with the formation of alcoholate and evolution of acetylene.

Rubidium Carbide (Rb_2C_2) is decomposed by water with evolution of acetylene. Dilute and concentrated hydrochloric acids readily decompose it, while nitric acid reacts explosively with it.

The carbide is not oxidized by dry oxygen at ordinary temperatures [64], but in a stream of fluorine, chlorine, bromine, or iodine it burns at room temperature. Liquid sulfur, phosphorus, arsenic, silicon, and boron react with Rb_2C_2 on slight heating. Carbon dioxide and sulfur dioxide, and also oxides of nitrogen, decompose it only on heating; CO is reduced by the carbide at 700-800°C to atmospheric carbon. The carbide reduces the oxides of copper, iron, and chromium.

On heating a mixture of rubidium carbide and calcium in vacuum, metallic rubidium is given off and deposited on the cold furnace walls in the form of a shining mirror.

Cesium Carbide (Cs_2C_2) is decomposed by water with the formation of cesium hydroxide and evolution of acetylene [64]:

$$Cs_2C_2 + 2H_2O = 2CsOH + C_2H_2.$$

It burns in concentrated nitric acid.

Halogens react explosively with it at room temperature. Cs_2C_2 reacts similarly with boron and amorphous sulfur.

Cesium carbide is a powerful reducing agent and on slight heating it reduces iron oxide.

Copper Carbide (Cu_2C_2) is practically insoluble in water. It reacts with a solution of KHC_2 in liquid ammonia in accordance with the equations [80]

$$Cu_2C_2 + 2KHC_2 = 2CuHC_2 + K_2C_2,$$
$$2CuHC_2 + 4KHC_2 = 2K_2[Cu(HC_2)_3].$$

Beryllium Carbide (Be_2C) readily burns in a stream of fluorine to form soluble beryllium fluorides [64]. It burns in chlorine at 700-800°C with the formation of volatile chlorides and precipitation of graphite. Bromine reacts with Be_2C only at high temperatures; iodine does not react with it at 800°C. Hydrogen fluoride reacts with Be_2C at 450°C. On passing from hydrogen fluoride to the chloride, bromide, and iodide, the temperature at which the reaction takes place increases, reaching 750°C for HI. The respective halogenides of beryllium are formed in this way.

Be_2C is oxidized by oxygen at 700–800°C, an oxide film being formed which protects the carbide from further oxidation. The carbide reacts with sulfur vapor at 1000°C to form beryllium sulfide. In contact with moist air at room temperature [88], Be_2C undergoes rapid hydrolysis to form $Be(OH)_2$.

The carbide is slowly decomposed on heating in concentrated hydrochloric and nitric acids. It reduces sulfuric acid to sulfurous acid. Hydrofluoric acid readily decomposes it.

A feature of beryllium carbide is its dissolution in water at room temperature with evolution of methane:

$$Be_2C + 4H_2O = CH_4 + 2Be(OH)_2.$$

A study of the reaction of Be_2C with silicon and nickel has been carried out in [95].

An investigation of the solubility of the carbide in molten copper—beryllium alloy [96] has shown that no reaction takes place that leads to decomposition of the carbide. At 1700°C the dissociation of Be_2C becomes significant in accordance with the equation

$$Be_2C_{solid} \rightleftarrows 2Be_{liq} + C_{graphite}$$

Hönigschmid [64] analyzed beryllium carbide by decomposing it by heating in a stream of pure dry chlorine. In this procedure volatile beryllium chloride was given off, from which the beryllium content was determined.

The beryllium was determined in [96] in the form of BeO by a gravimetric method after boiling a weighed sample of the carbide in sulfuric acid (1: 1).

The total carbon content was determined in [64] by weighing the insoluble residue left after treating a sample of the carbide with chlorine. This method is not very efficient, the most widely used method of determining the total carbon content being by the combustion of a sample in a stream of oxygen.

To determine the free carbon content, in [96] the residue left after decomposing Be_2C by prolonged boiling in sulfuric acid (1: 1) was burned. This method may lead to considerable errors in determining the free carbon content, since on prolonged boiling the graphite may be oxidized by the sulfuric acid. To determine the combined carbon content, the amount of methane liberated by the hydrolysis of Be_2C was determined by combustion.

In determining the free carbon content in [765], a sample of the carbide was decomposed by sulfuric acid (1: 1) on heating with a reflux condenser. The insoluble residue was burned in a stream of oxygen at 900°C for 1 h, the CO_2 evolved being absorbed by askarite.

The beryllium was determined by precipitation from an ammonium phosphate solution as NH_4BePO_4 which, after heating, was weighed in the form of $Be_2P_2O_7$.

Magnesium Carbides (Mg_3C_2 and MgC_2) are decomposed by water, forming methyl acetylene [64]. The Mg_3C_2 content was determined from the amount of allylene liberated on treating the carbide with water.

Calcium Carbide (CaC_2) is readily decomposed by water with evolution of acetylene

$$CaC_2 + 2H_2O = C_2H_2 + Ca(OH)_2.$$

The acetylene liberated in this way is absorbed in an ammoniacal solution of copper chloride and is perfectly pure [64]. The above reaction proceeds when there is excess water, i.e., when the calcium carbide is thrown into water. When the carbide is in excess, the following reaction occurs [116]:

$$CaC_2 + Ca(OH)_2 = 2CaO + C_2H_2.$$

The effect of the composition of the carbide, its grain size, and the temperature of the water on the decomposition rate of the carbide was investigated in [743], and it was found that the nearer the carbide was in composition to the calculated value, the higher was the decomposition rate. An increase in the particle size sharply reduces the decomposition rate owing to the decrease in contact surface between the carbide particles and water. Raising the temperature of the water also increases the decomposition rate (Fig. 70). It was found that, in the decomposition of the carbide by water, acetylene is not liberated uniformly with time. About 97% of the carbide is decomposed very rapidly, the remaining 3% more slow-

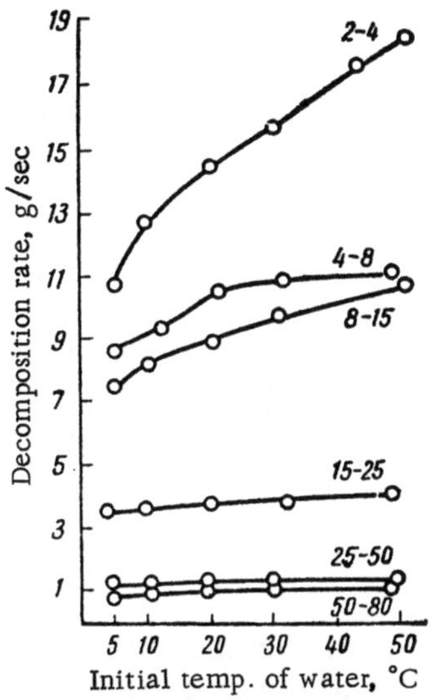

Fig. 70. Dependence of the decomposition rate of CaC_2 by water on temperature of the water and particle size of the carbide (the numbers on the curves represent millimeters).

TABLE 34. Dependence of the Decomposition Time of Calcium Carbide on Particle Size

Time, sec	Particle size of carbide, mm					
	2—4	4—8	8—15	15—25	25—50	50—80
Mean rapid decomp . . .	60	99	109	254	810	994
Mean total decomp . . .	255	259	304	488	930	1237
Ratio, %	23.5	38.3	35.9	52	87	80.5

ly. The ratio of the time of rapid decomposition to the time of complete decomposition depends on the particle size of the carbide (Table 34).

At temperatures below 250°C, water vapor reacts with CaC_2 to form $Ca(OH)_2$ and liberate C_2H_2; above 250°C, CaO and C_2H_2 are formed, while at 700-800°C, CaO, carbon, and hydrogen are formed [104].

TABLE 35. Rate Constants for the Nitration of Calcium Carbide

Accelerating additions	% of charge	K, μm/min at temperature, °C 850	900	950	1000	1050
Without additions	—	—	—	0.0055	0.02	0.075
$CaCN_2+C$	10	—	—	0.0065	0.025	0.085
$CaCl_2$	5	0.04	0.075	0.135	—	—
$CaCl_2$	2.5	0.018	0.043	0.085	0.19	—
$CaCl_2$	1	—	0.008	—	0.095	0.21
CaF_2	3	—	—	0.05	0.13	0.265
CaF_2	1.5	—	0.01	—	0.11	0.275
CaF_2	0.5	—	—	—	0.07	0.22
NaF	1	—	—	0.035	0.12	0.3
BaF_2	2.5	—	0.02	0.045	0.085	0.225
Na_2SiF_6	1	—	0.01	0.045	0.095	0.26

The reaction of calcium carbide with water vapor has been studied by x-ray and chemical analyses in [744]. It was shown that the reaction capacity of the various faces of CaC_2 crystals with respect to water vapor varies, decreasing in the order (100) > (001) > (110) > (101) > (111).

Dilute mineral acids and aqueous solutions of alkalis decompose CaC_2. Fuming sulfuric acid acts only slowly, but concentrated sulfuric acid rapidly decomposes the carbide; the carbide is hardly affected by fuming nitric acid at its boiling point.

Dry hydrogen does not react with calcium carbide at temperatures up to 2200°C. Above this temperature the following reaction takes place:

$$CaC_{2\,liq.} + H_2 \rightarrow Ca_{gas} + C_2H_2.$$

On heating, nitrogen reacts with the carbide in accordance with the equation

$$CaC_2 + N_2 = CaCN_2 + C,$$

this reaction being used industrially to produce calcium cyanamide.

The kinetics of nitration of CaC_2 powder is represented by the linear equation $\delta = k\tau$. The effect of accelerating additions and of the temperature of the process on the rate constant of the reaction is shown in Table 35 [745-747].

In a stream of ammonia at temperatures above 650°C the reaction

$$CaC_2 + 4NH_3 = CaCN_2 + NH_4CN + 4H_2$$

also takes place. In dry oxygen, CaC_2 is oxidized to calcium carbonate at 700-800°C. On heating in a CO atmosphere above 1600°C, a reaction proceeds in accordance with the equation

$$CaC_2 + CO = CaO + 3C.$$

Dry chlorine reacts with CaC_2 at 250°C to form $CaCl_2$ and carbon; sulfur vapor forms calcium sulfide CaS and CS_2; phosphorus forms the phosphide Ca_3P_2 and arsenic forms Ca_3As_2; in the last two cases, carbon separates as graphite.

Gaseous hydrogen chloride and hydrogen sulfide also react with the carbide to liberate acetylene. All these reactions are accompanied by the evolution of heat.

Sodium peroxide decomposes CaC_2 according to the reaction

$$7Na_2O_2 + 2CaC_2 = 2CaO + 4Na_2CO_3 + 3Na_2.$$

In the reaction of calcium carbide with potassium sulfocyanate [748], calcium sulfide, potassium cyanide, and graphite are formed

$$CaC_2 + 5KCNS = CaS_5 + 5KCN + 2C.$$

Organic chlorine-containing products react at 300-400°C as follows:

$$2CaC_2 + CCl_4 = 2CaCl_2 + 5C,$$

$$3CaC_2 + 2CHCl_3 = 3CaCl_2 + 8C + H_2.$$

Calcium carbide is a powerful reducing agent [749] and at high temperatures may completely reduce sulfates to sulfides:

$$\mathit{Me}SO_4 + 4CaC_2 = 4CaO + \mathit{Me}S + 8C,$$

in this reaction no evolution of CO is observed.

The oxides of lead, tin, zinc, iron, manganese, nickel, cobalt, chromium, molybdenum, and vanadium are reduced with the formation mainly of calcium alloys. On alloying with calcium carbide, the oxides of manganese, chromium, molybdenum, tungsten, titanium, and silicon are reduced to their respective carbides. In reacting with SiO_2, the silicates α-$CaSiO_3$, β-Ca_2SiO_4, and γ-Ca_2SiO_4, and also β-SiC, are formed, depending on the proportions of the components. Aluminum oxide is reduced by calcium carbide to metallic aluminum, aluminum carbide being formed on prolonged heating [116]. The mechanism of the reaction between calcium carbide and oxides has been described in [143-145].

Metallic calcium containing less than 1.3% impurities has been obtained by the reaction of calcium carbide with ferrotitanium at a temperature of 1450°C and a pressure of 66.6 N/m^2 (0.5 mm Hg) [140].

Experiments on the reaction between calcium carbide and boron have shown that at 1300°C the reaction leads to the formation of the hexaboride and the borocarbide. With excess of metallic calcium, the latter decomposes to form CaB_6 and CaC_2.

A borocarbide of composition CaC_4B_2 has been obtained by the reaction of calcium oxide with boric anhydride and carbon at 1900°C [142]; as a result of the reaction of calcium with carbon and boron at 1300°C, the borocarbide CaC_2B is formed [141]; on treatment with hydrochloric acid this compound decomposes with the evolution of hydrocarbons.

In the analysis of calcium carbide the calcium and combined carbon contents are determined. For this purpose a sample of the carbide is decomposed in water, and the calcium content is determined by a gravimetric method in the aqueous solution. The carbon content is determined from the quantity of acetylene evolved during decomposition of the carbide. The pressure and volume of the acetylene are measured with a manometer [750]. The error in the determination is 0.5%.

A rapid method for the direct determination of calcium carbide in slags has been proposed in [751]; in this method a sample is dissolved in dilute nitric acid saturated with potassium nitrate, and the acetylene liberated is burned in a Mars furnace. The carbon dioxide formed is absorbed by an alkali solution in a Wurtz —

Schtrohlein apparatus. The following method is used to determine the free carbon content [752]. A sample of the slag in a heated porcelain boat is moistened with a few drops of hydrochloric acid (1: 1) and dried. In this way all the carbide is decomposed with the evolution of acetylene, and the free carbon remaining in the slag is determined by the usual method. The calcium carbide content is determined from the difference

$$\% \, CaC_2 = (C_{total} - C_{free})\, 2.67,$$

where 2.67 is the factor for converting the C_{comb} content into CaC_2.

In the analysis of calcium carbide, the amount of calcium combined as carbide has been determined [126] from the quantity of hydrogen in the acetylene liberated on decomposing the sample with water. The amount of acetylene given off from the carbide under industrial conditions is determined from the volume of acetylene liberated, using the formula

$$V = \frac{(V_1 - 0.5)\, F \cdot 1000}{g}.$$

Here V is the volume of acetylene evolved during analysis, and

$$F = \frac{(P - w)\,(273 + 20)}{760\,(273 + t)},$$

where t is the temperature, P is the atmospheric pressure reduced to 0°C, w is the water vapor tension at temperature t, and g is the weight of the carbide sample, in g.

A method for the complete commercial analysis of calcium carbide has been developed in [823]; in this the amounts of CaC_2 and of the impurities SiO_2, Fe_2O_3, Al_2O_3, CaO, MgO, S, P, $CaCN_2$, and C_{free} are determined. The determination of CaC_2 is based on the measurement of the volume of acetylene evolved on decomposition by water. The residue left after the CaC_2 has been determined is evaporated with concentrated hydrochloric acid. The insoluble residue is melted with a mixture of soda and borax, and the silicon content is determined by a colorimetric method in the solution [824]. In aliquot parts of the filtrate after separating the insoluble residue, Fe_2O_3 is determined by a colorimetric method using sulfosalicylic acid, Al_2O_3 by the reaction with aluminum, and

MgO by means of titanium yellow dye [825]. The CaO content is determined by a complexonometric method, using murexide [826].

The free carbon content is determined by burning a sample of the carbide in a stream of oxygen at 800-900°C and subsequently titrating the CO_2 evolved with a solution of $Ca(OH)_2$. To determine the sulfur and phosphorus contents, a sample is decomposed in hydrochloric acid followed by the passage of a stream of nitrogen, H_2S, and PH_3; the sulfur is determined by an iodometric method after collecting the H_2S in a solution of calcium acetate, while the phosphorus is determined by a photocolorimetric method after collecting the PH_3 in a potassium permanganate solution.

Barium Carbide (BaC_2) is readily decomposed by water:

$$BaC_2 + 2H_2O = Ba(OH)_2 + H_2C_2.$$

BaC_2 is stable in dry air up to 600-700°C [64, 146]. Halogens readily decompose it: iodine vapor reacts at 122°, bromine vapor at 130°, and chlorine at 140°. The carbide is decomposed by fluorine on heating, with the formation of BaF_2.

At 700-800°C, barium carbide reacts with sulfur vapor to form barium sulfide and carbon bisulfide; it reacts similarly with selenium; phosphorus vapor forms the phosphide Ba_3P_2 and arsenic forms the arsenide.

Nitrogen reacts according to the equation

$$BaC_2 + N_2 = BaN_2C + C.$$

With CO,

$$BaC_2 + CO = BaO + 3C.$$

Heating BaC_2 with $Ba(OH)_2$ leads to the formation of benzene:

$$3BaC_2 + 3Ba(OH)_2 = 6BaO + C_6H_6.$$

Like calcium, barium reacts with carbon and boron at 1300°C to form the borocarbide BaC_2B; the borocarbide BaC_4B_2 is formed by the reaction of barium oxide with boric anhydride and carbon at 1900°C [141, 142]. The borocarbides are decomposed by dilute hydrochloric acid with the formation of liquid unsaturated open-chain hydrocarbons which polymerize into a solid amorphous substance.

Barium carbide is analyzed [64] in a similar way to calcium carbide. The carbide is decomposed in water, the barium is precipitated in the solution, and the amount of acetylene corresponding to the combined carbon content in the carbide liberated during decomposition is measured. In the Institute of Problems in Materials Science of the Academy of Sciences of the Ukrainian SSR, the contents of barium and of combined and free carbon are determined in the analysis of barium carbide. The barium content is determined, after decomposing a sample of the carbide in dilute (1:20) hydrochloric acid, by a complexonometric method by adding excess of trylon B and titrating the excess of trylon with a nickel chloride solution in the presence of murexide.

The determination of the combined carbon content is based on the decomposition of the carbide by water in a flask especially devised for the purpose and on the subsequent burning of the gases evolved in a stream of air. The amount of carbon dioxide formed in this way is determined by the ordinary absorption gas volumetric method.

To determine the free carbon content, a sample of the carbide is decomposed in water or dilute (1:20) hydrochloric acid. The free carbon in the form of an insoluble residue is filtered on a layer of heated asbestos and the residue is dried and burned in a stream of oxygen.

Scandium Carbide is easily decomposed by water to give the gaseous products CH_4 and C_2H_6 and also liquid and solid hydrocarbons. The chemical properties of ScC have scarcely been investigated. It oxidizes at 700-800°C in an oxygen atmosphere.

Yttrium Carbide is readily decomposed by water with the evolution of a mixture of hydrocarbons and hydrogen. The principal products of the decomposition are methane in the case of YC and Y_2C_3 (95 and 70%, respectively) and acetylene in the case of YC_2.

The carbides are considerably more stable in concentrated acids than in dilute ones, and more stable in H_2SO_4 than in HCl and HNO_3.

Yttrium dicarbide readily reacts with the halogens — with fluorine at room temperature and with the remainder at 700-800°C — and it easily burns in oxygen and in sulfur and selenium vapors [64]. In a study of the oxidation of yttrium carbide powders [185, 186], it has been shown that they decompose in air at room temperature mainly on account of the moisture in the air. In dry air and dry oxygen the dicarbide is stable at room temperature. A study of the oxidation of carbide powders at elevated temperatures by oxygen indicates that a large part of the carbides is oxidized at 500°C.

In the analysis of yttrium carbides the total and free carbon and yttrium contents are determined. The total carbon content is determined in the usual way by combustion in a stream of oxygen, followed by the absorption-volumetric determination of the carbon dioxide formed during the combustion. To determine the free carbon content, the carbide is decomposed in nitric acid (1:10) and the insoluble residue is burned in a stream of oxygen. The yttrium content is determined either by oxidation of a sample of the carbide to Y_2O_3 by combustion in a muffle furnace at 900-1000°C or by a volumetric method by titration with a solution of trylon B, using arsenazo as indicator [788].

Lanthanum Carbides react with water, as do the carbides of all the lanthanides.

The gaseous products of the decomposition of LaC_2 have been found to contain 53% C_2H_2, 15.6% H_2, 11.2% C_2H_4, and 20.2% C_2H_6 [8]. In the decomposition of the sesquicarbide the principal products are methane, 75%, and hydrogen, 25%.

An investigation of the gaseous products of decomposition of the dicarbide by hydrochloric acid and by a solution of iron chloride in hydrochloric acid has shown [790] that the composition of the products varies in relation to the reagent (Table 36).

Dilute mineral acids [64] readily decompose lanthanum dicarbide; concentrated sulfuric acid is reduced to sulfurous acid.

A study of the behavior of the dicarbide in various reagents indicates that it is almost completely soluble in mineral acids but is resistant to organic solutions.

TABLE 36. Composition of the Gaseous Products of Decomposition of LaC_2

Reagent	Content, %			
	C_2H_2	C_2H_4	CH_4	H_2
HCl	88.8	0.5	6.2	3.4
$FeCl_3$+HCl	96.7	0.2	1.2	1.0

TABLE 37. Composition of the Products of Decomposition of CeC_2 by Hydrochloric Acid and by $FeCl_3$ Solution in Hydrochloric Acid

Reagent	Content, %			
	C_2H_2	C_2H_4	CH_4	H_2
HCl	74	1.5	—	—
$FeCl_3$+HCl	94.7	Not det.	1.6	2.1

On heating in a stream of ammonia, LaC_2 is converted into the nitride; it readily reacts with the halogens, sulfur, and selenium to form halogenides, sulfides, and selenides; LaC_2 is decomposed by gaseous hydrogen chloride with the formation of $LaCl_3$ and the evolution of hydrocarbons; nitrogen and phosphorus do not react with it.

Although not affected by alkali solutions, the dicarbide is readily decomposed by molten alkalis and also by such oxidizing agents as potassium permanganate, chlorate, and nitrate.

The dicarbide is practically stable in dry air and oxygen at room temperature, but it rapidly decomposes in moist air.

In an oxygen atmosphere, dicarbide powder begins to oxidize at 500°C, oxidation proceeding rapidly at 800-900°C.

The nature of the oxidation is the same for LaC_2 and La_2C_3, but the oxidation rate of La_2C_3 is higher.

In the chemical analysis of lanthanum carbide [64], the total carbon content is determined by the ordinary combustion method and the free carbon content by decomposing the carbide with dilute nitric acid and weighing the insoluble residue. The lanthanum content has been determined by weighing the residue, representing La_2O_3, after combustion of the carbon [790].

The best method of determining the lanthanum content is a volumetric one, namely titration with a solution of trylon B, using arsenazo as indicator [788].

Cerium Carbide. When cerium dicarbide is decomposed by water, it gives rise to a mixture of hydrocarbons whose composition is variously reported by different authors (see Table 5).

The principal products of the decomposition of the sesquicarbide Ce_2C_3 by water are methane and hydrogen; in addition, ethylene and liquid and solid hydrocarbons are produced.

The composition of the gaseous products of the reaction of cerium dicarbide with hydrochloric acid and with a solution of iron chloride in hydrochloric acid is given in Table 37 [790].

The dicarbide is decomposed in dilute mineral acids; in concentrated sulfuric acid it is decomposed on heating, H_2SO_4 being reduced to H_2SO_3. It does not react with fluorine at room temperature, but on heating volatile fluorides are formed; with the remaining halogens the corresponding cerium halides are likewise formed on heating. Under the action of gaseous hydrogen chloride it decomposes at 650°C with the formation of cerium chloride. It behaves similarly with HI. Nitrogen and phosphorus do not react with CeC_2 at temperatures up to 800°C. With hydrogen sulfide it forms a mixture of cerium sulfide and graphite at 600°C.

The dicarbide is vigorously oxidized by oxygen at 800°C; the sesquicarbide is appreciably oxidized even at 600°C.

The analysis of cerium carbide is similar to that of lanthanum carbide.

Carbides of Praseodymium and Neodymium. Dilute acids act like water on the dicarbides of Pr and Nd. Concentrated acids act more slowly.

Both carbides readily react with the halogens to form the corresponding halides. The reaction with fluorine is slow at room temperature; with chlorine and iodine it takes place at 700-800°C, the reaction with chlorine being explosive. Nitrogen and gaseous ammonia decompose them at 1200°C with the formation of nitrides. Hydrogen sulfide reacts with the carbides at 700–800°C to form the sulfides.

In an oxygen atmosphere the carbides ignite at temperatures above 500°C. The dicarbides are fairly stable at room temperature in dry air and oxygen. The analysis of the carbides is carried out in a similar way to that of lanthanum carbide.

Samarium Carbide is easily decomposed by water with the evolution of a mixture of hydrocarbons. Mineral acids decompose SmC_2 the more readily the more dilute they are. In a stream of fluorine or chlorine the carbide decomposes on heating with the formation of the corresponding halide. With bromine and iodine the reaction occurs at 700–800°C. In oxygen SmC_2 burns at 400°C.

Thorium Carbides ThC_{1-x} and ThC_{2-x} are chemically unstable compounds and on heating in air they are easily oxidized to ThO_2. They are readily decomposed by water and dilute mineral acids. The compositions of the gaseous products of decomposition of the carbides by water are given in Table 5.

Uranium Carbides are chemically unstable compounds. A characteristic feature of their decomposition by water is that the decomposition products are gaseous, liquid, and solid hydrocarbons [228].

The monocarbide in the form of sintered specimens readily decomposes in boiling water [231]. A study of the effect of the impurities iron, silicon, tungsten, nitrogen, oxygen, and hydrogen on the decomposition of the monocarbide by water at 60°C has shown that iron and tungsten raise the decomposition rate, whereas hydrogen and oxygen lower it. On heating in water vapor, the monocarbide ignites and is converted into the black oxide. It reacts rapidly with water vapor at room temperature. In dilute mineral acids (HCl, H_2SO_4, HNO_3) the monocarbide reacts readily with the evolution of hydrogen, acetylene, ethylene, ethane, and other hydrocarbons. The action of concentrated acids is less vigorous. Phosphoric acid reacts slowly with the monocarbide at room temperature, but the reaction rate increases on heating.

The monocarbide is oxidized by CO_2 at 500°C [231, 774], the oxidation rate being constant over the range 500–830°C. A comparison of the reactions of the monocarbide and of metallic uranium with CO_2 has shown that the former is the more stable. Thus, at 500°C a carbide specimen oxidizes at the rate of 0.6% in 6 h, while for metallic uranium the rate is 6%. The monocarbide does not react with hydrogen until high temperatures are reached.

The monocarbide is oxidized by carbon dioxide according to the equation $UC + 2CO_2 = UO_2 + C + 2CO$. Above 500°C, the reaction $C + CO_2 = 2CO$ takes place, while at 1000°C, 50% of the carbon passes into the gaseous phase in the form of CO [774, 934]. The rate of oxidation of the monocarbide by carbon dioxide depends on the density of the specimens. At 80% of the theoretical density the change in weight as a result of oxidation is 140 $mg/cm^2 \cdot h$. The change in weight of a specimen with a density 90% of the theoretical value under the same oxidizing conditions is 10 $mg/cm^2 \cdot h$ [199].

In an investigation of the reaction of the monocarbide with CO (Fig. 71) [774], the dicarbide and dioxide of uranium were found by x rays in the reaction products. The authors conclude that the reaction proceeds in accordance with the equations

$$UC + 2CO = UO_2 + 3C,$$
$$UC + C = UC_2.$$

At room temperature the finely dispersed monocarbide is pyrophoric. The conditions under which the monocarbide and metallic uranium powders ignite have been investigated in [769]. The ignition temperature rises with increase in the rate of heating and the diameter of the powder particles. The temperature for uranium carbide is 40° higher than that for metallic uranium powder.

A study of the oxidation of the monocarbide by oxygen [774] has shown (Fig. 72) that the final oxidation product is U_3O_8 and that the reaction proceeds in accordance with the equation

$$3UC + 4O_2 = U_3O_8 + 3C.$$

At the same time the reaction $C + O_2 = CO_2$ takes place and 75% of the carbon passes into the gaseous phase.

The data of various authors on the oxidation of the monocarbide [774, 776, 777] show considerable discrepancies, probably ow-

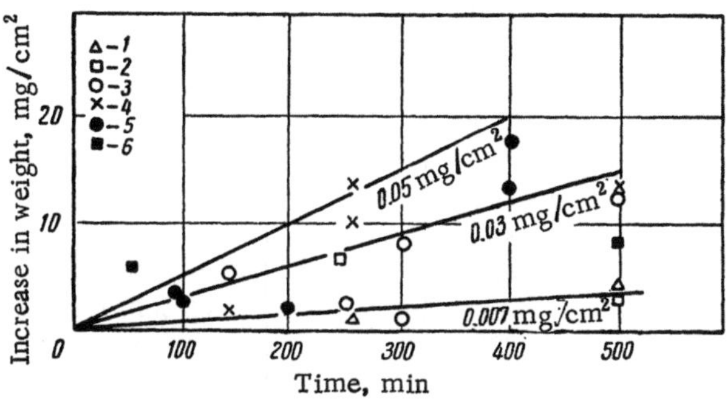

Fig. 71. Temperature dependence of the rate of reaction of uranium monocarbide with CO: 1) 400; 2) 450; 3) 500; 4) 600; 5) 750; 6) 1000°C.

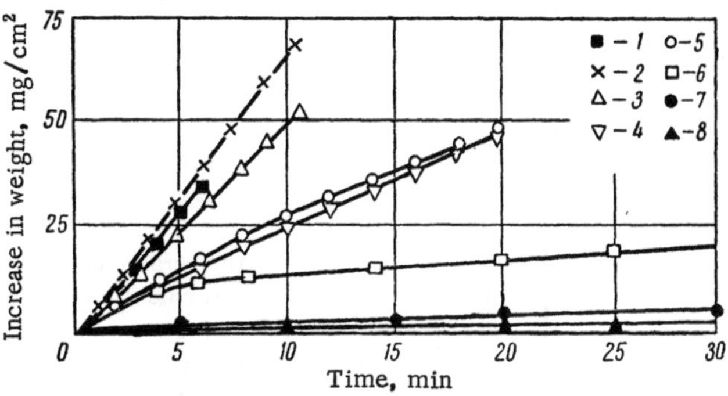

Fig. 72. Temperature dependence of the oxidation rate of uranium monocarbide: 1) 1000; 2) 850; 3) 700; 4) 550; 5) 500; 6) 450; 7) 400; 8) 350°C.

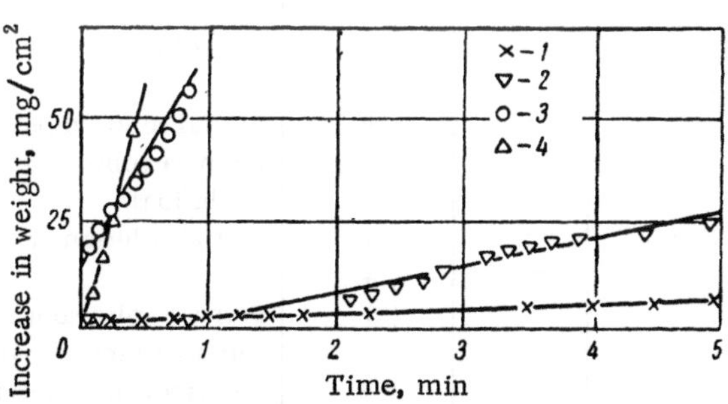

Fig. 73. Temperature dependence of the rate of reaction of uranium monocarbide with nitrogen: 1) 700; 2) 775; 3) 850; 4) 1000°C.

TABLE 38. The Reaction of Uranium Monocarbide with Some Metals and Carbides

Metal or carbide	Temp., °C	Nature of reaction*	Remarks
Sodium	< 500	–	
Sodium – potassium	800	–	
Beryllium	650	+	Under a load of 1.5 MN/m^2 (15 kgf/cm^2) for 12 h
Magnesium	870	–	For 330 h
Calcium	870	–	
Tin	–	–	
Lead	–	–	
Bismuth	600	–	
Aluminum	600	+	Forms UAl_3
Titanium	>1200	+	
Zirconium	1000	+	
Niobium	1000	+	
Tantalum	1800	–	
Chromium	1435	+	Forms a eutectic containing 34 ± 0.5% C
Molybdenum	–	–	
Iron	750	+	Forms ternary compound $UCFe_2$
	1170	+	Forms eutectic containing 51.5 ± 0.5% Fe
Stainless steel	1098	+	Forms a eutectic
Nickel	–	+	Forms the compound UNi_5
Rhodium, Iridium	–	+	Form ternary compounds of the type U_2MeC_2 [920]
Titanium carbide	–	+	Forms a 10% solution of titanium carbide in uranium monocarbide
Zirconium carbide	–	+	Forms a continuous series of solid solutions
Hafnium carbide	–	+	
Vanadium carbide	–	+	Partial solution observed
Niobium chromide	–	+	Forms a continuous series of solid solutions
Chromium carbide (Cr_3C_2)	–	+	Partial solution observed
Molybdenum carbide	–	+	
Tungsten carbide	–	+	Forms a solid solution containing up to 10 mol.% WC in UC
Thorium carbide	–	+	Forms a continuous series of solid solutions

* + reacts, – does not react.

ing to differences in the composition and density of the specimens investigated.

A study of the oxidation of monocarbide specimens prepared by various methods [777] led to the conclusion that the best results are obtained on specimens produced by arc melting. These specimens corresponded to the composition $UC_{1.02}$ and the density was very close to the theoretical value. On oxidation in pure dry oxygen in the range 550–800°C, an easily removed layer of composition U_3O_{8-x} was formed on the surface of the specimen. Between this layer and the surface of the carbide there was a layer of the oxide UO_2.

The oxidation obeys a linear law, the activation energy being 90.9 ± 4.2 kJ/mole (21.7 ± 1 kcal/mole).

The dependence of the reaction rate of the monocarbide with nitrogen is shown in Fig. 73.

Uranium monocarbide is stable in respect of organic heat carriers. No reaction was observed after the monocarbide had been held for 5 h at temperatures up to 350°C in diphenyl and triphenyl. On heating in glycerin at 100°C, the loss in weight was 50 mg/cm^2.

The results of an investigation into the reaction of uranium monocarbide with various metals are given in Table 38 [214, 231, 257, 258].

The sesquicarbide has been least studied. It does not react appreciably with water at 750°C [228]. It reacts vigorously with hydrochloric acid at 750°C; it reacts slowly with sulfuric and nitric acids at temperatures up to 170°C and does not react at all at these temperatures with acetic acid [231]. On heating in air to 600°C for 5.5 h, the sesquicarbide disintegrates into separate granules having a metallic luster. It reacts with iodine at 600°C to form uranium iodide.

The dicarbide is slowly decomposed by water at room temperature and, on heating, the decomposition rate increases considerably [231]. In the absence of air the green hydroxide is formed by the reaction with water; in the presence of air, gray-black products are formed. A mixture of hydrocarbons, one-third gaseous and two-thirds liquid and solid, is formed by the reaction [770].

The dicarbide reacts with water vapor at 700–800°C to form black uranium dioxide, the reaction rate obeying a linear law [231].

Dilute mineral acids (HCl, H_2SO_4, and HNO_3) rapidly decompose the dicarbide with the formation of yellow uranyl salt solutions [220] and a mixture of hydrocarbons. Concentrated hydrochloric, sulfuric, and phosphoric acids react slowly at room temperature, but rapidly on heating. The dicarbide is readily decomposed by alkalis. In air or in an oxygen atmosphere it ignites at 400°C with the formation of CO_2 and U_3O_8.

An investigation of the oxidation of uranium dicarbide [231] has led to the conclusion that up to 250°C the oxidation obeys a parabolic law. At 300°C the reaction proceeds with a great evolution of heat, as a result of which the temperature of the specimen is raised to 1000°C in the course of 1 min.

The reaction with nitrogen up to 700°C obeys the parabolic law. It proceeds very vigorously at temperatures above 1100°C. The carbide is completely converted into nitride on being maintained at 1180°C for 12 h. No reaction is observed with fluorine at room temperature, but on slight heating the reaction proceeds explosively. The reaction of the dicarbide with chlorine takes place at 350°C to form the volatile chloride. A mixture of UO_2 and UC_{2-x} partially reacts with chlorine at 600°C to form uranium tetrachloride and at 800–1000°C to form higher chlorides. The reaction with bromine begins at temperatures above 300°C, and at 900°C UBr_4 is formed and carbon is deposited. Iodine vapor reacts at 500°C to form UI_4. In a hydrogen sulfide atmosphere the dicarbide ignites at 600°C to form the sulfide; in an ammonia atmosphere it decomposes at 700–800°C.

On melting the dicarbide with potassium nitrate or perchlorate, potassium uranate is formed, while on heating a mixture of UC_{2-x}, UO_2, and NaCl to 1000°C the complex compound $Na_2UO_2Cl_4$ is formed.

In analyzing uranium carbides the uranium, total and free carbon, and nitrogen were determined in [225]. The uranium content was determined by the combustion of a 100-mg sample to form U_3O_8, followed by the appropriate conversion, or by a volumetric method by reducing the uranium by zinc amalgam to the tetravalent state and titrating the solution obtained with a $0.2N$ solution of potassium permanganate.

The total carbon content was determined by the combustion of a 0.1-g sample in a stream of air, followed by adsorption of the CO_2 evolved by soda lime. It was found in [227] that the sample is completely burned at 1260°C and no flux addition is necessary.

A more accurate method of determining the total carbon content has been proposed in [778]. Combustion is carried out for 15 min at 950-1050°C in a mixture of argon and oxygen. Then the rate of argon flow is gradually reduced, while the rate of oxygen flow is correspondingly increased and then maintained constant for 30 min. The amount of CO_2 liberated is determined from the increase in weight of absorption tubes containing NaOH deposited on asbestos and a layer of $Mg(ClO_4)_2$. At a carbon content of 4.64% and with a sample weighing 1.5 g, the accuracy of the determination is 0.03%.

Another method used for determining the total carbon content is based on the measurement of the thermal conductivity of the gaseous mixture of CO_2 and O_2 evolved in the combustion of a sample of uranium monocarbide in oxygen [945].

The sample is burned in oxygen at 1500°C, using a mixture of iron powder and tin as a flux. The gaseous products evolved pass into a special chamber, the reading of which is recorded either on a microammeter or on a special device calibrated in milligrams of carbon.

The free carbon content was determined by dissolving a 0.1-g sample in 50 ml of hydrochloric acid (1:1) by heating to the boiling point; then 1 ml of a 10% solution of hydrogen peroxide is added to the solution, which is boiled for 15 min. After cooling, the free carbon residue is filtered off, dried at 140-160°C, and weighed [225]. In [227] the free carbon content was determined by combustion of the residue left after treating a sample of the carbide with 1:1 hydrochloric acid and filtering on an asbestos layer, at 1260°C for 15 min.

The nitrogen content was determined in [225] by Dumas's method [754].

Titanium Carbide is resistant to the action of hydrochloric and sulfuric acids, but dissolves in aqua regia and a mixture of nitric and hydrofluoric acids [753]. The solubility of titani-

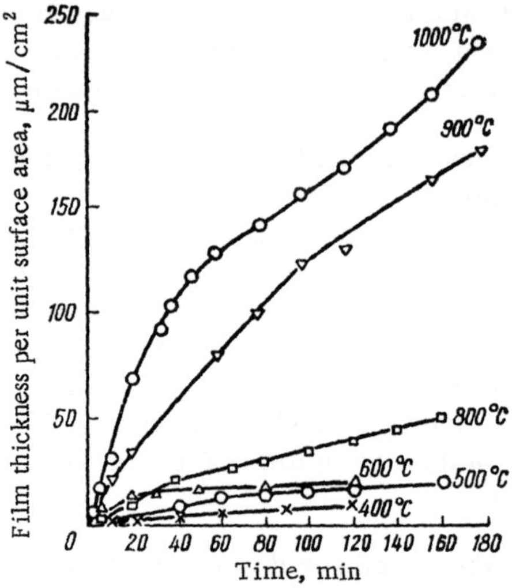

Fig. 74. Dependence of film thickness on oxidation time for TiC.

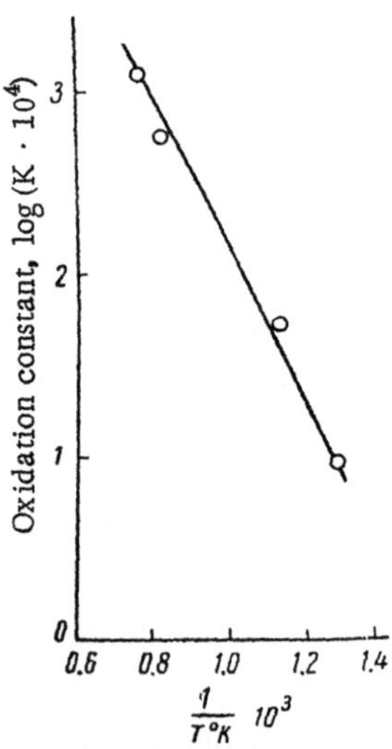

Fig. 75. Temperature dependence of the oxidation constants of TiC.

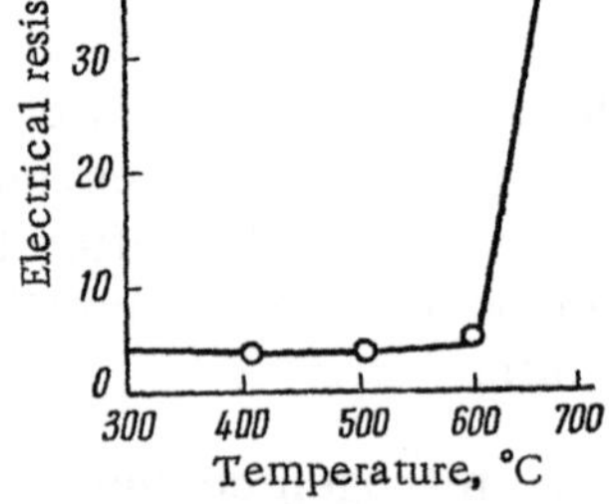

Fig. 76. Dependence of the electrical resistivity of TiC on oxidation temperature.

um carbide in mineral acids, mixtures of them, and alkali solutions [754, 755] are given in Tables 6 and 7.

Oxygen slowly oxidizes compact titanium carbide at 800°C. According to data in [756], a rutile phase is mainly formed during the oxidation, but the mechanism by which the carbon penetrates the TiO_2 layer is not explained in this paper.

A more thorough investigation of the oxidation of compact specimens of TiC in the range 400-1000°C was carried out in [757], the results being given in Figs. 74-76. Chemical, x-ray, and electron-diffraction studies of the oxide films have shown that in the first stage of oxidation a dense film of the solid solution TiC — TiO is formed on the specimens as a result of the introduction of oxygen atoms into points in the carbide lattice not occupied by carbon. Then oxygen diffuses through the solid-solution layer to form the higher oxides Ti_2O_3 and TiO_2. The apparent activation energy for the diffusion of oxygen ions through the solid-solution layer, as calculated by the authors of [757], is 179.7 kJ/mole (42,900 cal/mole). The moment at which growth of the higher oxide layer begins is indicated by a change in the electrical resistivity of the oxidized specimens (Fig. 76).

The resistance of titanium carbide specimens to oxidation in air [350] depends on the carbon content, and it is less in all cases than the resistance of molybdenum disilicide and zirconium diboride and is higher than the resistance of the carbide cemented with metal (Table 39).

Powdered titanium carbide is oxidized at 500-600°C with the formation of TiO_2 and CO_2. At 1200°C and above, the only gaseous product of the reaction is CO [758, 759].

The oxidation of titanium carbide in air at 1200°C and the conditions for the formation of protective coatings have been investigated in [137]. A coating of aluminum oxide, being porous, proved inadequate. As a result, thin coatings deposited on the aluminum oxide from semicolloidal solutions have been used. The solutions were prepared by mixing in various proportions aqueous solutions of boric acid and ammonium nitrate with a solution of silicic acid obtained by mixing 100 ml of freshly distilled silicoethyl ether with 45 ml of 85% ethyl alcohol, 15 ml of water, and 2 drops of sulfuric acid. The solution was applied by means of a

TABLE 39. Resistance of Titanium Carbide to Oxidation in Air at 1150°C

Material	Change in wt. (mg/cm · sec) in time, h					
	8	16	24	32	48	200
TiC_{1-x} 18.4% $C_{comb.}$ 1.3% C_{free}	6	7	7	9	Decomp.	—
TiC_{1-x} 13.3% $C_{comb.}$ 0.6% C_{free}	4	5	5	6	—	—
TiC_{1-x} +10% Co	10	18	Decomposed			
TiC_{1-x} +20% Co	8	12	—	—	—	—
TiC_{1-x} +10 % Ni	5	9	18	Decomposed		
$MoSi_2$	2	4	5	5	6	8
ZrB_2	0.5	1.2	2	3	3	4

TABLE 40. Thermodynamic Characteristics of the Reactions of Titanium Carbide with Vapors and Gases

Reaction	$-\Delta H_{298}$		$-\Delta S_{298}$		$\Delta C_{p_{298}}$	
	kJ/mole	kcal/mole	J/deg · mole	cal/deg · mole	J/deg · mole	cal/deg · mole
$TiC+2H_2O=TiO_2+C+2H_2$	276.9	66.1	84.6	20.2	209.5	5.0
$TiC+H_2O=TiO+C+H_2$	93.0	22.2	41.9	10.0	104.7	2.5
$TiC+{}^3/_2O_2=TiO_2+CO$	871.9	208.1	89.8	20.0	7.1	1.7
$TiC+O_2=TiO+CO$	446.2	106.5	3.35	—0.8	6.7	1.6
$TiC+2CO=TiO_2+3C$	540.1	128.9	353.6	84.4	—11.7	—2.8
$TiC+{}^1/_2N_2=TiN+C$	153.3	36.6	84.2	20.1	2.5	0.6
$TiC+4HCl=TiCl_4+2H_2+C$	209.5	50.0	153.3	36.6	10.9	2.6

paint sprayer to the surface of the titanium carbide specimen coated with aluminum oxide and heated to 600–800°C. On impact with the hot surface of the specimen, the solvents evaporated, leaving SiO_2, Al_2O_3, and B_2O_3 on the aluminum oxide layer. After being treated several times in this way, the specimen was heated at 1400°C for 5–10 min. As a result, the oxides SiO_2, Al_2O_3, and B_2O_3 formed a glass-like mass which filled the pores in the alumina layer. The optimum composition of the impregnated mixture

TABLE 41. Edge Angles of Wetting of Metals on a Titanium Carbide Surface

System	Temperature, °C	Edge angle θ	Atmosphere
Bi—TiC	300—700	138—118	Vacuum
Pb—TiC	400—1000	152—90	Argon
Cu—TiC	1100—1300	108—70	Vacuum
Ag—TiC	980	108	"
Al—TiC	700	118	"
Zn—TiC	550	120	"

is 60% SiO_2 and 40% B_2O_3. The use of this composition gives an aluminum oxide coating which protects the carbide from oxidation in air at 1200°C for 100 h and which will withstand 100 heat reversals in the temperature range 20-1200-20°C with cooling in a stream of air.

Water vapor oxidizes titanium carbide at high temperatures [760], and as a result the use of moist hydrogen in sintering TiC-base alloys leads to their decarburization.

Fluorine reacts with TiC at room temperature and the other halogens react at 400-700°C. Chlorine-containing compounds react vigorously with it, forming titanium chlorides, and in the presence of oxygen or sulfur forming the oxychloride or sulfochloride. Titanium carbide is resistant to nitrogen up to 1400-1600°C [761, 762], but above this temperature range the nitride is formed [762]. Nitrous oxide rapidly oxidizes it in accordance with the reaction

$$TiC + 3N_2O = TiO_2 + 3N_2 + CO.$$

Carbon monoxide does not react with the carbide [317]. Carbon dioxide oxidizes it in accordance with the equation $TiC + 3CO_2 = TiO_2 + 4CO$, the rate being described by the equation for a first-order reaction.

From an investigation of the resistance of TiC to gaseous media [763], it has been possible to determine certain thermodynamic constants for the reactions concerned (Table 40).

Such metals as titanium, zirconium, niobium, chromium, nickel, cobalt, and iron react vigorously with titanium carbide [277, 353-358].

From an analysis of the wettability of titanium carbide by various metals, it is possible to divide the latter into two groups according to their capacity for wetting the carbide surface [359]: (1) metals which react only slightly with the carbide surface and give large edge angles of wetting; and (2) metals which react vigorously with the carbide surface and spread completely over the surface, thoroughly wetting the framework of carbide grains. The results of the reaction of metals of the first group are given in Table 41.

Metals such as Fe, Co, Ni, Cr, Nb, Zr, and Ti belong to the second group.

The difference in electron structure of the metals which react with the carbide provides the basis for the division into the above two groups. Metals of the first group, which do not wet the carbides, are characterized by completely filled d-electron shells whereas metals of the second group — transition metals with unfilled d-shells — react vigorously with the carbides and thoroughly wet their surface.

The character of the reaction of titanium carbide with refractory metals and oxides is indicated in Table 42.

In the analysis of titanium carbide it is usual to determine the titanium content and the total and free carbon [754]. A sample is dissolved in mixtures of nitric and hydrofluoric acids, sulfuric and nitric acids, and hydrochloric and nitric acids. The titanium content is determined by a volumetric or gravimetric method.

The total carbon content is determined by the combustion of a 0.07-0.08-g sample of the carbide in a Mars furnace in a stream of oxygen at 1100-1200°C and subsequent determination of the amount of CO_2 liberated by the absorption-volumetric method.

The free carbon content is determined by the decomposition of an 0.3-0.5-g sample of the carbide in a mixture of nitric and hydrofluoric acids. In this procedure the carbide is dissolved, while the free carbon is left in the form of an insoluble residue. This is filtered off by means of a layer of heated asbestos; the residue is then washed, transferred to a porcelain boat, dried, and burned in a Mars furnace in a stream of oxygen, the amount of CO_2 given off being determined by the absorption-gas-volumetric method.

TABLE 42. Character of the Reaction of Titanium Carbide with Some Metals and Oxides

Metal (oxide)	Temp., °C	Nature of the reaction	References
Niobium	2000	Forms solid solution. Activation energy of diffusion of Nb into TiC = 352 kJ/mole (84 kcal/mole)	[363]
Tantalum	2000	Forms a TiC — TaC solid solution containing about 50 mol.% TaC	[360]
Chromium	2000	Forms a solid solution containing 7% Cr. The hardness of the carbide increases.	[361]
Molybdenum	2000	Forms a solid solution containing 20-22% Mo. The hardness of the carbide decreases.	[361]
Tungsten	2000	Forms a solid solution. The activation energy of diffusion of W into TiC = 481.8 ± 33.5 kJ/mole (115 ± 8 kcal/mole).	[361, 362]
Rhenium	2540	Small solubility of Re in TiC (5-6 wt.%) is observed.	[923]
Cobalt	1550	Reacts	[877]
Nickel	1500	Reacts	[877]
Carbon steel	1620	Reacts	[877]
Cast iron	1520	Reacts	[877]
Cryolite	1050	Reacts slightly	[877]
Magnesium oxide	1925	The reaction $2MgO + TiC = 2Mg + CO + TiO$ occurs. TiO dissolves in excess TiC. Formation of Mg_2TiO_4 and partial reduction of MgO to MgC_2 are possible; decomposition $3MgC_2 \rightleftharpoons Mg_3C_2 + 4C = 3Mg + 6C$ proceeds subsequently.	[364, 365]
Silicon oxide	—	Forms solid solution Ti(C, O) and SiO	[277]
Aluminum oxide	—	No reaction	[366]

In some cases the nitrogen content in titanium carbide is determined by a colorimetric method, using thymol and hypobromite [764].

Zirconium Carbide is a chemically stable compound which is practically insoluble in mineral acids at room temperature. It is resistant to dilute and concentrated hydrochloric acid on heating [25]. Tables 6 and 7 give data on the resistance of zirconium carbide to various mineral acids, to mixtures of them, and to alkali solutions [754].

TABLE 43. Thermodynamic Data for the Reaction of Zirconium Carbide with Some Gases

Reaction	$-\Delta H_{298}$		$-\Delta S_{298}$		$\Delta C_{p_{298}}$	
	kJ/mole	kcal/mole	J/deg · mole	cal/deg · mole	J/deg · mole	cal/deg · mole
$ZrC+2H_2O=ZrO_2+2H_2+$ $+C$	473.5	113.0	95.5	22.8	15.6	4.2
$ZrC+{}^3/_2O_2=ZrO_2+CO$	1068	255	94.7	22.6	3.8	0.9
$ZrC+2CO=ZrO_2+3CO$	736.6	175.8	363.7	86.8	—15.1	—3.6
$ZrC+{}^1/_2N_2=ZrN+C$	227.9	54.4	86.7	20.7	—3.8	—0.9
$ZrC+4HCl=ZrCl_4+2H_2+$ $+C$	478.9	114.3	330.2	78.8	31.0	7.4

The carbide does not react with water vapor at temperatures up to 600–700°C. It is readily decomposed by halogens: by fluorine at room temperature, by chlorine at 250°C, by bromine at 300°C, and by iodine at about 400°C [64].

The reaction of zirconium carbide with iodine has been investigated in [767]; the reaction proceeds fairly rapidly at 700–800°C with the formation of ZrI_2.

Heating a compact specimen in air at 1150°C for 48 h leads to the disintegration of the specimen [350].

The carbide reacts with nitrogen above 1500°C to form zirconium nitride. It does not react with ammonia.

Some thermodynamic data for the reaction of zirconium carbide with a number of gases are given in Table 43 [763].

Such oxidizing agents as potassium chlorate, nitrate, and permanganate in the molten state readily decompose zirconium carbide [766].

A study of the reaction of the carbide with NH_4HF_2 [768] has shown that the reaction begins at room temperature and proceeds fairly rapidly with the formation of $(NH_4)_3ZrF_7$ and the evolution of the gaseous products H_2, NH_3, C_2H_2, and CH_4.

TABLE 44. Character of the Reaction of Zirconium Carbide with Metals and Oxides

Metal	Reaction temp., °C	Nature of the reaction	References
Niobium	1600	Forms ZrC — NbC solid solution	[360, 394, 395]
Tantalum	2200	Forms tantalum carbide and a solution of Zr in Ta having hardness greater than that of Ta	[360, 394]
Molybdenum	1900	Forms a new phase similar in hardness to zirconium	[360, 394]
Tungsten	2200	Forms a new phase of great hardness exceeding that of tungsten	[360, 394]
Rhenium	–	Very slight solubility of Re in ZrC (< 0.8 at.%) is observed	[922]
Tin, bismuth, zinc, lead, cadmium	350-550	No reaction is observed during 10-min contact	[360, 394]
Magnesium oxide	–	Partial reduction of MgO with the formation of MgC and the solid solution Zr(C, O)	[365]
Zirconium oxide	1800	Forms an oxycarbide of the type ZrC_xO_{1-x}	

Table 44 shows the nature of the reaction of zirconium carbide with some metals and oxides.

In the analysis of zirconium carbide it is usual to determine the zirconium content and the total and free carbon. To determine the zirconium content [303, 304] the carbide is melted with a mixture of KOH and KNO_3 and leached with water; the solution is acidified with hydrochloric acid, and the zirconium is precipitated with ammonia. Because zirconium carbide is soluble in concentrated sulfuric acid and also in a mixture of nitric and hydrofluoric acids, it is more convenient to dissolve it than to melt it. After solution, the zirconium can be determined by a volumetric, gravimetric, or amperometric method [754].

On heating the carbide in a stream of chlorine, the metal volatilizes in the form of the chloride; the residue, containing all the carbon, is weighed and then it is treated with fuming nitric acid; in this way the free carbon is left as graphite [64].

In later researches [303, 304, 754], the total and free carbon contents have been determined by the procedures used for titanium carbide.

The amount of unreacted ZrO_2 has also been determined in [303]. For this purpose a sample was treated for 1 h at 140-155°C in a mixture of 100 ml H_2SO_4 (1: 1) and 10 ml chromic acid (540 g Cr_2O_3 in 300 ml H_2O). The carbide and free carbon were dissolved in this way. A 75% solution was neutralized with caustic soda, diluted with water, and filtered. The filter with the ZrO_2 residue was burned and weighed.

The total oxygen content was determined from the difference $100 - (Zr_{total} + C_{total})$; the zirconium combined in the form of carbide was determined from the difference $Zr_{total} - Zr$ in ZrO_2; the amount of oxygen present in the carbide was determined from the difference $O_{total} - O$ in ZrO_2.

Hafnium Carbide is resistant to hydrochloric, dilute sulfuric, and phosphoric acids and to alkali solutions.

Its solubility in various reagents [755] is given in Tables 6 and 7. The character of the reaction of hafnium carbide with some metals and oxides is given in Table 45.

In the analysis of hafnium carbide the hafnium content and the total and free carbon were determined in [754].

The hafnium content was determined by a gravimetric cupferron method or by heating a sample at 900-1000°C.

The total and free carbon contents were determined by methods similar to those used for titanium and zirconium carbides.

Vanadium Carbide is not decomposed by sulfuric and hydrochloric acids, although it is decomposed by nitric acid at room temperature.

Oxygen oxidizes it at 800°C; chlorine reacts with the carbide at temperatures above 500°C to form volatile liquid vanadium chloride [64]; ammonia and nitrogen form vanadium nitride at 700-800°C. Gaseous hydrogen chloride, water vapor, and hydrogen sulfide do not react with the carbide at 700-800°C.

The vanadium content of the carbide has been determined by combustion of a sample at 950°C. Because vanadium pentoxide

TABLE 45. Character of the Reaction of Hafnium Carbide with Some Metals

Metal	Temp., °C	Nature of the reaction	References
Zinc	550	Does not react	[877]
Cadmium	450	Does not react	
Aluminum	1000	Reacts	
Silicon	1500	Reacts	
Tin	350	Does not react	
Lead	450	Does not react	
Niobium	1600	Forms a HfC — NbC solid solution having great hardness and a high melting point	[360]
Tantalum	1600	Reacts slowly	
Chromium	1900	Does not react	[877]
Molybdenum	1600	Reacts vigorously with the formation of a solid solution of Mo in the carbide	[360]
Tungsten	1600	Reacts slowly with the formation of a new phase having a hardness of 33,320-34,300 MN/m^2 (3400-3500 kgf/mm^2) and with an increase in hardness of the tungsten	
Cobalt	1550	Does not react	[877]
Nickel	1500	Does not react	[365]
Magnesium oxide	2000	Does not react	

melts at 715°C, wets the walls of the boat, and may pass over to the tube of the Mars furnace, the authors filled the boat with finely ground and calcined quartz powder. The molten V_2O_5 wetted the surface of the quartz and was retained in the boat. At 1100°C the oxide was observed to lose weight, a sign of its decomposition.

The combustion of vanadium carbide is known to lead frequently to the formation of a mixture of oxides of various compositions, and in some cases the combustion is accompanied by ejection of the product. Finally, the use of the quartz powder on whose surface the V_2O_5 melts is a source of great error in determining the vanadium content by the method described. As a result, the classical volumetric method of determining vanadium in the carbide, the decomposition of which presents no difficulty, is undoubtedly more expedient [754, 793, 794].

A micromethod of determining the total carbon content has been suggested in [795, 796]; in this method the gases evolved by combustion of the carbide in oxygen are collected in a special absorption bottle and determined by titration with a $Ba(OH)_2$ solution, the excess of which is then titrated back with hydrochloric acid.

To avoid possible ignition of the carbide during combustion in a stream of oxygen and the passage of CO, copper oxide is placed in the end of the combustion tube. Fairly satisfactory results for the total carbon content are given by the combustion of a sample in a stream of oxygen at 900-1000°C, followed by the absorption-gas-volumetric determination of the amount of CO_2 liberated. A refinement of the method of determining the free carbon content [792] has shown that it is most conveniently dissolved in a 2 : 1 nitric acid solution. A 0.2-0.3-g sample of VC_{1-x} dissolves in it in a few minutes on raising the temperature to the boiling point.

Niobium Carbide is a chemically stable compound whose properties are given in Tables 6 and 7.

The carbide oxidizes in air at 800-900°C, and it reacts with nitrogen and ammonia at high temperatures to form nitrides [64]. Whereas the chemical properties of the monocarbide have been studied to some extent, those of the lower carbide Nb_2C_{1-x} are almost completely unknown.

In an investigation of the oxidation in air of powder and pieces of Nb_2C_{1-x} of size up to 3 mm at 300°C [132], a new phase was found in the reaction products; it has a rhombic structure and its composition corresponds approximately to $NbC_{0.4}$. In this work it was found that $a_{comb} = \sqrt{3}a_{hex}$, $b_{comb} \approx c_{comb}$, and $c_{comb} \approx a_{hex}$.

The character of the reaction of niobium monocarbide with refractory metals and oxides is given in Table 46.

Niobium carbide is analyzed in the same way as the carbides of titanium and zirconium.

Tantalum Carbide is one of the most chemically stable carbides. Its solubility in various reagents is given in Tables 6 and 7.

TABLE 46. Character of the Reaction of Niobium Carbide with Refractory Metals and Oxides

Metal	Temp., °C	Nature of the reaction	References
Niobium	1600	Forms Nb_2C	[394, 360, 899]
Tantalum	1800	Forms a layer having a microhardness of 18,620 MN/m^2 (1900 kgf/mm^2), which is similar to that of the solid solution $Nb_2C - Ta_2C$	
	2000	Forms two layers: one adjacent to the tantalum with a microhardness of 18,530 MN/m^2 (1890 kgf/mm^2) and an outer layer with a microhardness of about 34,000 MN/m^2 (3470 kgf/mm^2)	
Molybdenum	1800	Reacts with the formation of a porous layer whose microhardness increases with rise in temperature until it attains values approaching those of Mo_2C and MoC. Microhardness of molybdenum also increases as a result of the solution of carbon in it	
Tungsten	2000	Forms WC	
Magnesium oxide	1800	Reacts vigorously with the formation of a phase with a microhardness of 21,440 MN/m^2 (2189 kgf/mm^2), presumably the solid solution Nb(C, O) and a soft phase adjacent to it with a microhardness of 1460 MN/m^2 (149 kgf/mm^2)	[365]
Niobium oxide	1200	Forms lower oxides and the lower carbide Nb_2C; metallic niobium is formed at 1600-1700°C	[918]

Heating the carbide in hydrogen [472] leads to its decarburization if the concentration of the carbon-containing phase (CH_4, %) in the hydrogen falls below the following values:

Temperature, °C	In hydrogen	In mixture of 25% N_2 and 75% H_2
2300	$^1/_8$—$^1/_4$	$^1/_8$—$^1/_4$
2510	$^1/_8$—$^1/_4$	$^1/_8$—$^1/_4$
2770	$^1/_8$—$^1/_4$	$^1/_4$—$^1/_2$
3180	$^1/_4$—$^1/_2$	$^1/_2$

TABLE 47. Thermodynamic Data for Reactions between Tantalum Carbide and Some Gases and Vapors

Reaction	$-\Delta H_{298}$		$-\Delta S_{298}$		ΔC_{p298}	
	kJ/mole	kcal/mole	J/deg · mole	cal/deg · mole	J/deg · mole	cal/deg · mole
$2TaC + 7H_2O = Ta_2O_5 + 2CO + 7H_2$	251	60.0	−47.3	−11.3	88.4	21.1
$2TaC + {}^7/_2O_2 = Ta_2O_5 + 2CO$	1944	464.0	263.5	62.9	18.0	4.3
$2TaC + 5CO = Ta_2O_5 + 7C$	1172	279.8	891.2	212.7	−26.0	−6.2
$TaC + {}^1/_2N_2 = TaN + C$	90.1	21.5	104.3	24.9	1.26	0.3
$TaC + 5HCl = TaCl_5 + {}^5/_2H_2 + C$	−61.6	−14.7	533.0	127.2	−18.8	−4.5

Nitrogen and a mixture of 15% N_2 and 85% Ar do not react with the carbide at temperatures up to 2700°C. Oxygen oxidizes it to Ta_2O_5 at 800°C.

Table 47 contains thermodynamic data relating to reactions between tantalum carbide and some gaseous reagents [763].

An investigation of the reaction between tungsten and tantalum carbide [362] by measuring the radiation intensity from a layer of radioactive tungsten applied to a thin section of TaC by diffusion annealing has shown that the activation energy of tungsten diffusion is 595 ± 92 kJ/mole (142 ± 22 kcal/mole), while the diffusion coefficients are $2.7 \cdot 10^{-12}$ cm²/sec at 2130°C, $4.2 \cdot 10^{-12}$ cm²/sec at 2170°C, $9.0 \cdot 10^{-12}$ cm²/sec at 2230°C, and $17 \cdot 10^{-12}$ cm²/sec at 2300°C.

The character of the reaction between tantalum carbide and some refractory metals and oxides is given in Table 48.

The tantalum and free carbon contents are determined by the method already described for the carbides of titanium, zirconium, etc.

A micromethod of determining the total carbon content in tantalum carbide has been proposed in [795]. In this method a sample undergoes combustion in a stream of oxygen at 900–950°C

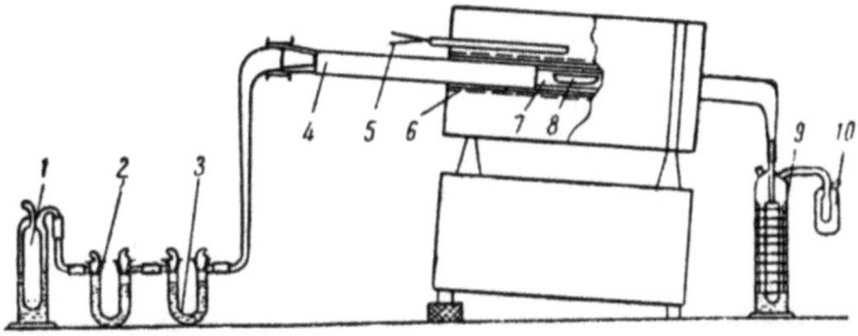

Fig. 77. Diagram of apparatus for determining small quantities of carbon in tantalum carbide. 1) Wash bottle containing H_2SO_4; 2) tube containing BaO; 3) tube containing P_2O_5; 4) quartz tube; 5) thermocouple; 6) Mars furnace; 7) protective quartz tube; 8) porcelain boat; 9) wash bottle containing $Ba(OH)_2$; 10) control flask containing $Ba(OH)_2$.

TABLE 48. The Character of the Reaction between Tantalum Carbide and Some Refractory Metals and Oxides

Metal	Temp., °C	Nature of the reaction	References
Niobium	1200	Forms a solid solution	[360, 899]
Tantalum	2200	Forms Ta_2C	
Molybdenum	2200	Forms very hard new phases, presumably ternary Ta – Mo – C phases	
Tungsten	2200	The same	
Magnesium oxide	2200	Does not react	[365]
Tantalum oxide	–	Forms metallic tantalum	[481]

in the apparatus shown in Fig. 77; the temperature is raised to 600°C in the absence of oxygen and from 600 to 900–950°C in a stream of oxygen which is passed for 30–40 min at a rate of 3 ml/min. The CO_2 liberated after the furnace has been switched on is swept by oxygen into a wash bottle filled with $Ba(OH)_2$ titration solution. The bottle is disconnected from the system and its contents titrated with hydrochloric acid in the presence of phenolphthalein in a stream of inert gas used to agitate the solution. The wash bottle consists of individual bells placed on the tube through which the gas passes and having at its upper end a hole through which the gas escapes. The tube is enlarged at its lower end and is filled with rings consisting of sections of narrow glass tube.

Chromium Carbides are characterized by their high resistance to corrosion and oxidation.

The carbide Cr_3C_2 is resistant to all mineral acids, but is oxidized by molten potassium nitrate and sodium peroxide [483]. The carbide is decomposed by boiling concentrated perchloric acid with the evolution of acetylene.

The solubility of the carbide Cr_7C_3 in hydrochloric acid has been studied in [57, 797, 798]. It was shown that alcoholic solutions of hydrochloric acid act more strongly than do aqueous solutions of the acid on the carbide; this is due to the fact that the carbide has a certain tendency to passivation in aqueous solutions but loses the passivity in alcoholic solutions.

The solubility of the carbide is also affected by the concentration of hydrochloric acid; dissolution proceeding more completely in a 10% solution of the acid than in a 1% solution.

The resistance of chromium carbides to various acids, to mixtures of acids, and to alkali solutions increases in the order $Cr_{23}C_6 - Cr_7C_3 - Cr_3C_2$; this is due to the complexity of the structural elements consisting of carbon atoms [805, 806]. The least resistant carbide ($Cr_{23}C_6$) is characterized by the isolated location of the carbon atoms in the elementary cell, whereas the most resistant carbide (Cr_3C_2) is characterized by the presence of zigzag chains of carbon atoms.

In the presence of oxidizing agents the chromium carbides are increasingly resistant to decomposition in the order $Cr_3C_2 - Cr_7C_3 - Cr_{23}C_6$; this is connected with the formation of a passive film of Cr_2O_3 on the surface. This film is most resistant on the lowest carbide $Cr_{23}C_6$, which has the chromium lattice with carbon atoms located at the lattice points. The carbides Cr_7C_3 and Cr_3C_2 have more complex structures and differ in their properties from metallic chromium.

The electrochemical behavior of the carbides in sulfuric and hydrochloric acid solutions has been studied by determining the curves of current density plotted against electrode potential [925]. The results showed that Cr_3C_2 is resistant to oxidizing and reducing action in acids and that it has a high anodic polarizability.

According to numerous investigations [800-806], chromium carbides are the most resistant to high-temperature oxidation of all the carbides of transition metals belonging to groups IV, V, and VI of the periodic system.

Below are given comparative data on the oxidation resistance of some carbides [800] (temperature 900°C, time 1 h):

Carbide	Cr_3C_2	Cr_7C_3	$Cr_{23}C_6$	TiC	Nb(Ta)C	ZrC	VC	WC
Loss in wt., g/m^3.	0.66	1.1	1.1	12.1	205	460	435	1140

Investigation of the oxidation resistance of chromium carbides in the form of powders and compact specimens at temperatures from 400 to 1100°C [802] has shown that the oxidation of all three carbides in powder form begins at about 700°C.

Compact specimens of Cr_3C_2 and $Cr_{23}C_6$ remain practically unoxidized up to 1000°C, but compact Cr_7C_3 oxidizes appreciably even at 700°C.

The thermodynamic characteristics of certain reactions between Cr_3C_2 and some gases and vapors have been calculated in [763] (Table 49).

In the chemical analysis of chromium carbides, the chromium content and the total and free carbon are determined.

In [483] the total carbon content was determined after decomposing a sample of the carbide by melting with potassium nitrate or sodium peroxide. The mixtures obtained were treated with sulfuric acid, and the carbon dioxide liberated as a result was absorbed in weighed U-tubes containing soda lime. Melting was carried out in quartz flasks, which withstood only four melts. This method is not very effective, as it is essential to take into account the CO_2 content in the sodium peroxide, which readily absorbs it; this is a complicating factor which is accompanied by large errors.

The carbon content was determined in [485] by combustion after addition of aluminum powder to the carbide sample. The large amount of heat liberated by oxidation of the aluminum promotes complete decomposition of the carbide. The aluminum carbide, which is formed to some extent in this process, is oxidized by the addition of lead peroxide. Thus, to a 0.4-g sample of carbide is added 2 g of a mixture consisting of one part aluminum powder, one

TABLE 49. Thermodynamic Data Relating to the Reaction of Cr_3C_2 with Some Gases and Vapors

Reaction	$-\Delta H_{298}$		$-\Delta S_{298}$		$\Delta C_{p_{298}}$	
	kJ/mole	kcal/mole	J/mole·deg	cal/mole·mole	J/mole·keg	cal/mole·deg
$2Cr_3C_2+13H_2O = 3Cr_2O_3+4CO+13H_2$	481.8	115.000	—108.5	—25.9	191.1	45.6
$2Cr_3C_2+{}^{13}/_2O_2 = 3Cr_2O_3+4CO$	3630	866.400	468.9	111.9	60.3	14.4
$2Cr_3C_2+9CO = 3Cr_2O_3+13C$	2192	523.200	2053	490.1	—21.4	—5.1
$Cr_3C_2+{}^3/_2N_2 = 3CrN+2C$	271.5	64.800	165.5	39.5	41.5	9.9
$Cr_3C_2+9HCl = 3CrCl_3+2C+{}^9/_2H_2$	370.8	88.500	802.8	191.6	63.3	15.1

part aluminum oxide, and 10 parts lead peroxide. Combustion was carried out at 750°C.

The best conditions for the complete combustion of chromium carbide in oxygen consist of combustion for 25–30 min at 1150–1200°C, using copper oxide as flux [487, 804]. A check on this method showed that the error in determination is about 1%.

A study of the chemical resistance of chromium carbide to various reagents has shown that it is impossible to choose a reagent which will decompose only the carbide or only the free carbon. Hence the methods usually employed for determining the free carbon content cannot be used in the present case.

In [483] the difference in solubility of the chromium carbides in hydrochloric acid has been used to determine the free carbon content in the carbides. The residue insoluble in HCl was filtered off and weighed, and the chromium and carbon contents in it were determined. Since the residue consists of Cr_3C_2 and free carbon, the amount of carbon combined in the form of Cr_3C_2 could be determined from the chromium content. The free carbon content was determined from the difference between the total carbon content in the insoluble residue and the amount of carbon combined in Cr_3C_2.

This method is very long and not very accurate; hence, it is not a suitable method for determining small amounts of free carbon.

The free carbon content was determined in [807, 808] by combustion of a sample of the carbide in a stream of oxygen at about 600°C. In this way the free carbon was completely oxidized, but no oxidation of the carbide occurred.

In [807] a 0.3-0.5-g sample of the carbide was burned in a stream of oxygen at 600 ± 20°C for 30 min. The amount of CO_2 liberated was determined by the absorption-volumetric method.

In [808] a 0.5- 1-g sample was burned in a stream of oxygen at 560°C for 1 h. The amount of CO_2 liberated was determined from the increase in weight when it was collected in absorption tubes. In this paper a method was proposed for determining the free and combined carbon contents on a single sample. The carbide sample was burned at 560°C for 1 h; then the tube was changed, the temperature raised to 1300-1450°C, and combustion continued for 90 min.

A method has also been proposed for determining the free carbon content by combustion of a carbide sample at 540°C with absorption of the CO_2 liberated by alkali, followed by the conductometric determination of the amount of carbon absorbed [809].

All these methods are fairly effective; they do not differ in principle and can be successfully used to determine the free carbon content in chromium carbides.

The chromium content in the carbides was determined in [483] by decomposing a sample by melting it with potassium nitrate or sodium peroxide. The mixture obtained was treated with sulfuric acid, and the acid solution of bichromate thus formed was alkalized with caustic soda solution containing hydrogen peroxide. Iron and aluminum present in the carbide as impurities were precipitated as hydroxides and filtered off. The chromium content was determined in the filtrate by an iodometric method after addition of hydrochloric acid [794, 810].

The possibility of determining the chromium content by decomposing a sample of the carbide with boiling perchloric acid has been indicated in [497]; the determination is completed by use of the persulfate — silver method [793, 794].

TABLE 50. Thermodynamic Data for the Reaction of Molybdenum Carbide with Some Vapors and Gases

Reaction	$-\Delta H_{298}$		$-\Delta S_{298}$		$\Delta C_{p_{298}}$	
	kcal/mole	kcal/mole	J/mole·deg	cal/mole·deg	J/mole·deg	cal/mole·deg
$Mo_2C+7H_2O= =2MoO_3+7H_2+CO$	73.7	17.600	135.7	32.4	71.6	17.1
$Mo_2C+{}^7/_2O_2= =2MoO_3+CO$	1586.3	378.600	446.6	106.6	1.26	0.3
$Mo_2C+{}^1/_2N_2= =Mo_2N+C$	87.1	20.800	85.1	20.3	23.9	5.7

A check on this method has shown that during this treatment of the sample chromium losses occur owing to the volatility of the chromyl chloride which is formed to some extent.

The most effective method of determining the chromium content is by the persulfate — silver method after decomposition of the sample by melting with sodium peroxide or caustic soda.

Molybdenum Carbide. Nonoxidizing acids do not decompose Mo_2C_{1-x}, but it is dissolved by nitric acid and mixtures of it with other acids.

Data on the solubility of Mo_2C_{1-x} in various reagents [754] are given in Table 6. Under the action of water on its surface, a blue layer of oxides is gradually formed on the carbide.

On heating in air, the carbide powder begins to oxidize at 500°C, and at 800°C it is completely oxidized with the formation of MoO_3.

Fluorine reacts with the carbide at room temperature, but chlorine reacts only at high temperatures.

Thermodynamic data (Table 50) have been calculated from a study of the reaction of the carbide with certain vapors and gases.

No reaction of any kind is observed after 15 min contact of a compact carbide specimen with MoO_2 at 1700°C, with SiO_2 at 1300°C [539], with liquid copper at 1200°C, with open-hearth furnace slag at 1500°C, with Fe_2SO_4 at 1350°C, or with $CaSiO_3$ at 1600°C; on the other hand it does react with liquid nickel, cobalt, manganese, aluminum, and copper [529]. Molybdenum carbide reacts with tantalum at 1700-1900°C [552]. The reaction with molybdenum takes place at 1400°C with the formation of a Mo_2C — Mo solid solution, while the carbide reacts with tungsten at 1500°C to form a WC — Mo solid solution [899].

In [811] the carbide has been observed to react vigorously with an alkaline solution of $K_3[Fe(CN)_6]$ even at 20°C, CO being liberated in accordance with the reaction

$$Mo_2C + 14\,[Fe(CN)_6]^{3-} + 18\,(OH) = 2MoO_4^{2-} + \\ + 14\,[Fe(CN)_6]^{4-} + 9H_2O + CO.$$

The author has proposed a method of determining combined carbon which is based on the ready solubility of molybdenum carbide in an alkaline solution of $K_3[Fe(CN)_6]$ and the almost complete insolubility of free carbon in the solution. The procedure is as follows: a sample of the carbide is dissolved in 10 ml of a solution consisting of 105 g/liter $K_3[Fe(CN)_6]$ and 57 g/liter NaOH. The solution is made up to 300 ml with water and then 30 ml of a zinc sulfate solution (200 g/liter) are added, followed by drops of hydrochloric acid until the pH of the solution reaches 1.5. Then 20 ml of KI solution (50 g/liter) are added and titrated with a 0.05 *N* solution of $Na_2S_2O_3$.

The carbon content is determined from the equation

$$\%C = 4.53\frac{V}{E} - 60.1,$$

where V is the consumption of $[Fe(CN)_6]^{3-}$ in mg, and E is the weight of the sample in mg.

The determination of the combined carbon content by the volumetric method agrees well with the results of the combustion method, the difference being ± 0.06%.

The most widely used method for determining the total carbon content is the combustion method described above on page 226. The free carbon content is determined by the method usually employed for the majority of carbides, namely, solution of a sample

TABLE 51. Thermodynamic Data for the Reactions of WC with Water Vapor, Oxygen, and Hydrogen

Reaction	$-\Delta H_{298}$, kJ/mole	$-\Delta H_{298}$, kcal/mole	$-\Delta S_{298}$, J/deg·mole	$-\Delta S_{298}$, cal/deg·mole	$\Delta C_{p_{298}}$, J/deg·mole	$\Delta C_{p_{298}}$, cal/mole·deg
$WC+4H_2O=$ $=WO_3+4H_2+CO$	−54.5	−13.000	−15.9	−3.8	62.0	14.8
$WC+3H_2O=$ $=WO_3+CH_4+H_2$	152.1	36.300	198.6	47.4	15.1	3.6
$WC+H_2O=W+$ $+CO+H_2$	−166.8	−39.800	−142.0	−33.9	19.3	4.6
$WC+2O_2=$ $=WO_3+CO$	914.2	218.200	161.7	38.6	21.8	5.2
$WC+2H_2=W+$ $+CH_4$	39.8	9.500	72.9	17.4	−27.6	−6.6

in nitric acid 1:1 followed by combustion of the insoluble residue in oxygen [754].

A colorimetric method of determining the free carbon content in molybdenum carbide has been put forward in [812]. The method is based on the fact that the carbide does not adsorb such dyes as bromthymol blue from aqueous solutions.

Below are given the comparative results of determining the free carbon content by the colorimetric and gas volumetric methods; they show the good agreement between the data obtained by the two methods:

	I	II
Gas-volumetric method .	4.18	2.10
Colorimetric method . . .	4.26	2.09
Relative error, %	(+0.9)	(−0.5)

Tungsten Monocarbide. The behavior of WC in various chemically aggressive media is indicated in Table 6.

From a study of the reaction of WC with water vapor, hydrogen, and oxygen [763], it has been possible to calculate certain thermodynamic characteristics of the reactions (Table 51).

On heating in air or oxygen, WC slowly oxidizes to WO_3.

A study of the oxidation of WC [814] has shown that at 700°C the oxidation obeys a linear law and proceeds to completion in accordance with the reaction $WC + 2.5O_2 = WO_3 + CO_2$. The oxide film contains only WO_3. The degree of oxidation of WC is considerably greater than that of metallic tungsten owing to the rupture of the oxide film by the CO_2 formed.

Thermogravimetric analysis has shown [813] that fine powder begins to oxidize rapidly in air at 500-520°C. The rapid oxidation of coarsely crystalline WC begins at 565°C.

Fluorine reacts explosively with WC at room temperature, while chlorine begins to react only at 600-800°C [754].

An investigation of the electrochemical properties of WC, the determination of the static potentials, and the study of the dependence of the static potentials on the pH of the reagents and on the composition of the electrolyte have been carried out [815] by a compensation method with reference to the saturated calomel electrode, using an LP-58 potentiometer. The dependence of the static potential of WC on the reagent is given below:

Electrolyte	Static potential, V
0.001 *N* HCl (pH = 3)	+617
0.001 *N* HCl + 1% NaCl	+278
0.001 *N* HCl + 6.6% NaCl	+507
0.001 *N* HCl + 1% NaCl and 10% citric acid	+647
5% NaOH	−293
5% NaOH + 1% NaCl	−270
5% NaOH + 10% NaCl	−348
5% NaOH + 1% citric acid	−213
5% NaOH + 10% citric acid	−58

An investigation of the reaction of WC with molybdenum and tungsten has shown that in the first case two phases, Mo_2C and a Mo — W — C solid solution, are formed at 1850°C and in the second case W_2C is formed at 1900°C.

The diffusion of tungsten into WC has been investigated in [362] by measuring the radiation intensity of WC specimens to which a layer of radioactive tungsten had been applied by prolonged diffusion annealing at 2130, 2170, 2230, and 2300°C. The activation energy is 578.2 ±79.6 kJ/mole (138 ± 19 kcal/mole). The diffusion coëfficient, in cm^2/sec, is $1.9 \cdot 10^{-12}$ at 2130°C, $5.4 \cdot 10^{-12}$ at 2170°C, $7.8 \cdot 10^{-12}$ at 2230°C, and $16 \cdot 10^{-9}$ at 2300°C.

A study of the reaction of WC with the oxides V_2O_5, Nb_2O_5, and Ta_2O_5 in vacuum [931] has shown that in all cases the reaction begins at 900°C and attains a maximum at 1400°C. In the process solid solutions of tungsten are formed with the respective metals.

The solution of WC in an alkaline solution of $K_3[Fe(CN)_6]$ takes place rapidly in accordance with the reaction [811]

$$WC + 9\,[Fe\,(CN)_6]^{3-} + 11\,(OH)^- = WO_4^{2-} + \\ + 9\,[Fe\,(CN)_6]^{4-} + 5.5H_2O + {}^1/_2\,CO + {}^1/_2\,CO_2.$$

The solubility of WC in the $K_3[Fe(CN)_6]$ solution has been utilized by the author for the volumetric determination of the WC content or the combined carbon by the volumetric method (see page 241).

The WC content is calculated from the equation

$$WC = 35.4\frac{V}{E} - 245.1\%,$$

$$C = 2.17\frac{V}{E} - 15.0\%,$$

where V is the consumption of $[Fe(CN)_6]^{3-}$ in mg, and E is the weight of the sample in mg.

A comparison is made below of the results of determining the combined carbon content in WC by the volumetric and combustion methods, in %:

Volumetric method	3.12	6.02	4.87	1.42
Combustion method	3.27±0.03	5.89±0.04	4.3±0.06	1.5±0.05

The free carbon content can also be determined by the colorimetric method described in [812], which is based on the fact that methyl orange and methyl violet do not absorb WC, but do absorb free carbon. The procedure is similar to that described for determining the free carbon content in MoC. The results of determining C_{free} by this method are given in Table 52.

TABLE 52. Results of Determining C_{free} by the Colorimetric Method

Carbide	C_{free}, % content	C_{free}, % found	Relative error, %	Remarks
WC	1.83	1.76	−3.7	Absorption by methyl orange
WC	1.10	1.05	−4.5	
W_2C	1.64	1.56	−4.8	
WC	1.83	1.78	−2.7	Absorption by methyl violet
WC	0.77	0.74	−4.0	
WC	1.40	1.37	−2.1	
WC	1.76	1.73	−1.7	
W_2C	2.13	2.20	+3.2	

Manganese Carbide. The chemical properties of manganese carbide have not been adequately studied.

The carbide Mn_3C is rapidly decomposed by water with the evolution of a mixture of methane and hydrogen in the approximate proportions 51% CH_4 and 49% H_2. The carbide is also readily decomposed by dilute mineral acids.

The gaseous products liberated by the treatment of manganese—carbon specimens containing various amounts of carbon by hydrochloric acid (1:1) have been analyzed in [588].

The authors showed that on the dissolution of metallic manganese only hydrogen is liberated; Mn_4C on decomposition liberates, in addition to hydrogen (71%) and methane (24.2%), up to 5% ethane, while Mn_3C liberates 62% hydrogen, 27.9% CH_4, 9.3% C_2H_6, and 2% C_3H_8; i.e., the number of constituents on the gaseous mixture increases with rise in the carbon content, thus indicating the increasingly complex nature of the bond.

Fluorine reacts with Mn_3C at room temperature to form manganese fluoride, and chlorine reacts with it on heating.

Manganese carbide easily ignites on slight heating in an oxygen atmosphere and also in an atmosphere of NO and NO_2. It forms manganese nitride in ammonia at 700–800°C [64].

The manganese and combined carbon contents were determined in the analysis of manganese carbides in [588]. Manganese was precipitated in the form of the sulfide by passing hydrogen sulfide through a solution obtained by dissolving a sample of the carbide in dilute hydrochloric acid; then the sulfide precipitate was heated until Mn_3O_4 was formed. The combined carbon content was determined by combustion of the gases liberated during the dissolution of the sample in dilute hydrochloric acid.

Iron Carbide (Fe_3C) is insoluble in water at room temperature, but its solubility increases with rise in temperature above 150°C. Dilute mineral acids dissolve Fe_3C above 80°C; in the process, liquid and solid hydrocarbons, mainly propylene and butylene, are formed in addition to gaseous hydrocarbons. These products result from the polymerization of CH_2 groups [785, 791].

Finely divided Fe_3C powder ignites in an oxygen atmosphere at 150°C; in chlorine the reaction takes place at 100°C, in bromine and iodine at 700–800°C, and in sulfur vapor at 500°C.

The chemical properties of iron—carbon alloys and steels have mainly been reported in the literature, but the properties of the individual carbide phases have scarcely been studied at all.

Cobalt Carbide Co_3C, like iron carbide, is decomposed by dilute mineral acids on heating, the products being polymerization products of methylene [785, 791]. The total carbon content of Co_3C has been determined in [153] by combustion of a sample in a Corliss apparatus for 5 h, while the free carbon content has been determined by dissolving a sample in 30% HNO_3 followed by filtration of the insoluble free carbon.

Nickel Carbide Ni_3C is dissolved by concentrated and dilute mineral acids.

In studying the products obtained by dissolving the carbide in hydrochloric acid [827], a sample was placed in a flask which was heated on a water bath. Air was expelled from the flask by means of carbon dioxide and acid was added drop by drop. The gases evolved were collected in an azotometer containing a concentrated alkaline solution. The gaseous decomposition products consist mainly of hydrocarbons and hydrogen, the composition of the products being variable. The difference in composition of the reaction gases is due in the author's opinion to such processes as the formation of methylene groups, hydrogenation of methylene to methane, polymerization of methylene groups to acetylene, and hydrogenation of ethylene to ethane.

TABLE 53. Chemical Properties of Boron Carbide [848, 849]

Reagent	Temp., °C	Treatment time, h	Insoluble residue, %
HCl (conc.)	20	24	98.0
HCl (conc.)	115	1	98.0
HCl (1 : 1)	110	4	97.6
$HCl+HClO_4$	116	4	97.5
H_2SO_4 (conc.)	20	24	98.0
H_2SO_4 (conc.)	280	1	98.0
H_2SO_4 (1 : 1)	120	4	97.0
$H_2SO_4+HNO_3$	130	4	91.2
H_2SO_4+HF	280	1	91.5
HNO_3 (conc.)	20	24	97.0
HNO_3 (conc.)	120	1	97.0
HNO_3 (1 : 1)	110	4	96.9
HNO_3+HF	116	2	90.8
$HClO_4$ (conc.)	20	24	98.0
$HClO_4$ (conc.)	120	1	98.0
$HClO_4$ (1 : 1)	115	4	97.8
$KIO_3+H_3PO_4$	Boiling	3	89.7
$H_2SO_4+KMnO_4$	"	3	81.1
$H_2SO_4+K_2Cr_2O_7$	"	3	100.0
$H_2SO_4+H_3PO_4+CrO_3$	"	3	87.8
NaOH 1%	20	40	99.0
NaOH 1%	100	1	99.0
NaOH 1%	100	2	98.5
NaOH 5%	20	40	98.3
NaOH 6%	20	40	98.6
NaOH 8%	100	2	98.5
NaOH 12%	20	40	99.0
NaOH 12%	100	2	99.0
NaOH 25%	20	40	99.1
NaOH 10% $+H_2O_2$	100	1	98.0
NaOH 10% $+Br_2$	100	1	99.6

The chemical analysis of nickel carbide has been carried out by determining the nickel content and the total and free carbon.

To determine the free carbon content, a sample of the carbide was decomposed by boiling with concentrated hydrochloric acid for 30 min. The free carbon left as an insoluble residue was filtered off, dried, and weighed. Nickel was precipitated from the solution by dimethylglyoxime. To determine the total carbon content, a sample was burned in a stream of oxygen. As the samples ignited immediately on being placed in a hot furnace, they were put into a cold furnace and the temperature gradually raised.

Boron Carbide is one of the most chemically stable compounds. According to the data in [682, 772], it does not dissolve in mineral acids, in mixtures of them, or in alkaline solutions of various concentrations.

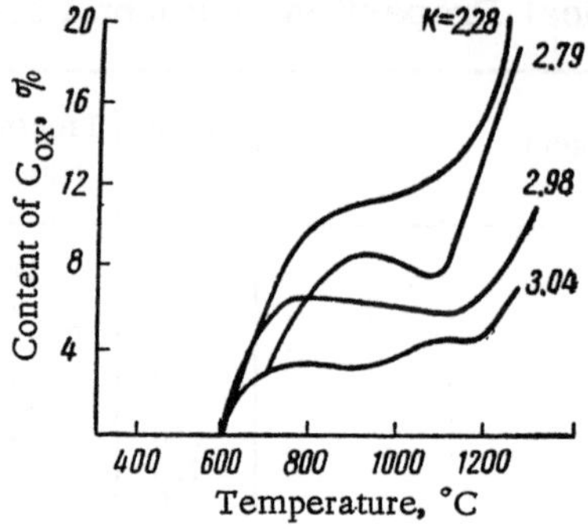

Fig. 78. Oxidation curves for boron carbide specimens of various compositions: K = B_{comb}/C_{comb}.

TABLE 54. Results of the Treatment of B_4C with a Mixture of $H_2SO_4HNO_3 + HClO_4 + K_2Cr_2O_7$ (10 ml : 10 ml : 5 ml : 20 mg)

Treatment time, min	Carbon content, %		Boron content, %		Iron content, %	
	before treat.	after treat.	before treat	after treat.	before treat.	after treat.
15	26.3	23.7	70	75	1.1	0.24
23	26.3	22.4	70	77	1.1	1.13
25	26.14	21.4	69.8	77.5	0.23	0.07

Data are reported in [847] which show that boron carbide is completely dissolved in 6 h by a mixture of 2 ml H_2SO_4 + 0.5 ml HNO_3.

The most thorough investigation of the chemical properties of boron carbide was made in [848, 849]. The results, given in Table 53, confirm the high chemical resistance of B_4C.

The absence from boron carbide of a large amount of free carbon is an important factor in its use.

The separation of the free carbon is based on the difference in oxidizing properties of the carbide and the free carbon. The following reagents have been used as oxidizing agents for the free carbon: 30% solution of hydrogen peroxide [846], a solution of chromic anhydride or potassium bichromate in sulfuric acid [850, 851], a mixture of concentrated sulfuric, nitric, and perchloric acids with potassium bichromate [852].

The most effective reagent for the oxidation of carbon black is the mixture proposed in [852, 853] (Tables 53 and 54).

Sulfur, phosphorus, and nitrogen do not react with boron carbide at 1200°C; chlorine reacts with it at 1000°C with the formation of BCl_3 and the separation of graphite; bromine and iodine do not react with it.

Oxygen does not oxidize boron carbide at temperatures up to 500°C, but at 1000°C the carbide ignites with the formation of CO_2 and B_2O_3 [754]. The oxidation of the carbide has been studied in [854] on powders of various grain sizes at 820 and 1800°C. It was found that the free carbon oxidizes considerably more rapidly than does the carbide, the rate of oxidation of the latter being constant. On this basis the authors have developed a method for determining the free carbon content by combustion of a sample in an oxygen atmosphere until a constant value of the ratio $\Delta C/\Delta t$ is reached.

An investigation of the oxidation of boron carbide powder in [853] has shown that oxidation begins at 600°C; the oxidation rate increases sharply with rise in temperature to 800-1000°C; it then becomes slower before again sharply increasing at 1200-1300°C (Fig. 78). At all temperatures oxidation proceeds for a certain period of time, after which it ceases. This is probably due to the formation on the B_4C particles of a B_2O_3 film which prevents further oxidation not only of the carbide but also of the carbon. This conclusion is supported by the fact that the most rapid oxidation of boron carbide occurs at temperatures at which B_2O_3 is very volatile. The greater the carbon content in the boron carbide, the higher is the oxidation rate.

In the reaction of boron carbide powder with tantalum platelets in the temperature range 1200-1400°C in a vacuum of 39.9 kN/m^2 ($3 \cdot 10^{-4}$ mm Hg), a coating is formed which is up to 4 μm thick and consists of TaB_2, TaB, and Ta_2B [739].

In the analysis of boron carbide the total and free boron, total and free carbon, and boric anhydride are determined [754].

To determine the total boron content a sample of the carbide is decomposed by melting with a mixture of sodium peroxide and caustic soda in iron crucibles, by sintering with barium carbonate or calcium oxide in nickel crucibles, or by melting with soda in platinum crucibles. After bringing the sample into solution, the boron content is determined by a volumetric method, namely, by titration of the boromannitic acid with a caustic soda solution in the presence of phenolphthalein. The determination of the free boron (including B_2O_3) is based on its dissolution in a mixture of nitric acid and perhydrol [914], in which boron carbide is not decomposed. Similarly, the determination of the boric anhydride content is based on its decomposition by water. In both cases the boron content is determined in the filtrate by a volumetric method.

The total carbon content is determined by combustion of a sample in a stream of oxygen at 1200°C, using copper oxide, lead oxides, or lead [721] as flux, followed by the absorption gas volumetric or gravimetric determination of the amount of CO_2 evolved. As indicated above, the determination of the free carbon content is based on the difference in the rate and character of oxidation of the carbide and free carbon [854]. Considering the oxidation of boron carbide and of free carbon as an additive process, it should be borne in mind that the oxidation rate of B_4C is constant ($\Delta C/\Delta t = K$) and that the oxidation rate of free carbon is higher than that of boron carbide. As a result, combustion is carried out at 1000°C until the ratio $\Delta C/\Delta t$ becomes constant. The free carbon content is then calculated from the formula

$$C_{free} = C - \frac{\Delta C}{\Delta t}t,$$

where C is the total amount of carbon consumed at time t.

Aluminum Carbide. A characteristic feature of aluminum carbide is its reaction with water to liberate methane

$$Al_4C_3 + 12H_2O = 3CH_4 + 4Al(OH)_3,$$

which takes place slowly at room temperature.

The carbide is decomposed very slowly by concentrated hydrochloric and nitric acids, but rapidly by the dilute acids. According to [164], methane and up to 8% hydrogen are liberated when the carbide is treated with hydrochloric acid. Concentrated sulfuric acid is reduced by aluminum carbide to SO_2. Molten caustic soda readily decomposes Al_4C_3 at 300°C.

Oxygen oxidizes the carbide at 700-800°C only on the surface, as an oxide film is formed which protects it from further oxidation. Chlorine decomposes the carbide at 700-800°C with the formation of aluminum chloride. Bromine reacts at 700°C, but iodine does not react at 800-900°C. Sulfur decomposes Al_4C_3 at 700-800°C with the evolution of heat and the formation of aluminum sulfide. Nitrogen and phosphorus have no action on the carbide.

Aluminum carbide is a powerful reducing agent; it reacts with the oxides of copper and lead at 700-800°C, and iron oxide is reduced by it. The reaction

$$Al_4C_3 + Al_2O_3 = 6Al + 3CO$$

takes place at high temperatures.

On heating, finely divided aluminum carbide reacts with $KMnO_4$, $K_2Cr_2O_7$, PbO, Pb_3O_4, KNO_3, and $KClO_3$.

The analysis of aluminum carbide has been carried out by two methods in [64].

1. The carbide was decomposed by the prolonged action of dilute hydrochloric acid. After the insoluble residue had been filtered off, the solution containing aluminum chloride was evaporated to dryness and the residue calcined to give Al_2O_3.

2. The carbide was decomposed by melting with caustic alkalies in a silver crucible; the mixture was dissolved in water, acidified with hydrochloric acid, and the aluminum precipitated by hyposulfite as hydroxide, which was then filtered off. The carbon content was determined by measuring the amount of methane evolved during the slow dissolution of a carbide sample in water.

In the analysis of aluminum carbide in [169], the volume of methane liberated during decomposition of a carbide sample was also measured.

Aluminum carbide was analyzed in [164] by heating a sample with 4 *N* solution of sulfuric acid. The methane evolved was burned in a tube over copper oxide, and the carbon dioxide formed was absorbed in tubes containing caustic soda which were then weighed. To absorb the hydrogen sulfide and sulfur dioxide which were evolved to a certain extent as impurities, a tube containing chromium trioxide was installed. Heating was carried on for 2 h. As a check, heating was continued for a further 1 h after weighing. Al_2O_3 and free carbon remain in the insoluble residue, which is filtered through a Gooch crucible, dried at 300°C, and weighed. The insoluble residue, representing 5.9% of the total, contained 3.7% C_{free} and 2.2% Al_2O_3. In one part of the filtrate from the insoluble residue the aluminum content was determined, and in another part nitrogen was determined by treatment with NaCl and titration of the distillate with hydrochloric acid.

Silicon Carbide is a chemically resistant compound. It is not decomposed by mineral acids or mixtures of them, except for phosphoric acid, nor is it affected by alkaline solutions [754, 828-831]. Phosphoric acid (density 1.75 g/cm³) decomposes silicon carbide at 200-250°C with the formation of SiO, CO_2, H_2, and CH_4 [702].

Data on the solubility of SiC in a mixture of nitric and hydrofluoric acids are conflicting. Some solubility of SiC in a mixture of these acids is indicated in [694], but according to [832] only 0.01% of the carbide passes into solution after six treatments with a mixture of HNO_3 + HF. These data are confirmed by our own investigations [833]. The discrepancy is evidently due to a certain difference in solubility between the various forms of SiC.

A detailed study of the effect of the crystalline structure of silicon carbide on its chemical resistance has been made in [929]. It was shown there that irrespective of grain size, α-SiC is not decomposed by hydrochloric acid or by mixtures of it with nitric and sulfuric acids; β-SiC is decomposed by these acids, the extent of the decomposition increasing with reduction in the temperature at which the β-SiC is produced.

Silicon carbide is stable in a hydrogen atmosphere up to high temperatures. In its reaction with nitrogen above 1400°C, the nitride Si_3N_4 is formed [702]. SiC readily reacts with fluorine; with chlorine it forms $SiCl_4$, and C at 900-1000°C and $SiCl_4$ and CCl_4 at 1100-1200°C; carbon monoxide reacts above 1250°C with SiC, depositing free carbon on its surface; sulfur reacts appreciably above 1100°C. SiC begins to react with water vapor at 1300-1400°C, and at 1775-1800°C the reaction

TABLE 55. Change in Free Energy in the Oxidation of Silicon Carbide

Reaction	Free energy at temperature, °K							
	298,16		1000		1500		2000	
	kJ/mole	kcal/mole	kJ/mole	kcal/mole	kJ/mole	kcal/mole	kJ/mole	kcal/mole
$SiC+2O_2=SiO_2+CO_2$	−1172	−279.8	−1049.6	−250.5	−972.0	−232.0	−892.0	−212.9
$SiC+{}^3/_2O_2=SiO_2+CO$	−912.2	−217.7	—	—	—	—	—	—
$SiC+O_2=SiO_2+C$	−779.7	−186.1	−573.2	−156.8	−576.5	−137.6	−489.0	−116.7
$SiC+{}^3/_2O_2=SiO+CO_2$	−451.0	−110.1	—	—	—	—	—	—
$SiC+O_2=SiO+CO$	−205.3	− 49.0	−333.9	− 79.7	−422.0	−100.7	−505.3	−120.6
$SiC+{}^1/_2O_2=SiO+C$	− 70.0	− 16.7	−134.9	− 32.2	−179.3	− 42.8	−212.8	− 50.8
$SiC+O_2=Si+CO_2$	−345.7	− 82.5	−351.1	− 83.8	−357.8	− 85.4	−367.9	− 87.8
$SiC+{}^1/_2O_2=Si+CO$	− 88.4	− 21.1	−157.5	− 37.6	−204.9	− 48.9	−257.3	− 61.4
$SiC+2CO=SiO_2+3C$	−509.0	−121.5	−258.9	− 61.8	− 91.3	− 21.8	+ 95.9	+ 22.9
$SiC+CO=SiO+2C$	+ 65.4	+ 15.6	+ 64.1	+ 15.3	+ 63.3	+ 15.1	+ 66.6	+ 15.9
$SiC+4H_2O=SiO_2+CO_2+4H_2$	−246.8	− 58.9	−282.0	− 67.3	−316.8	− 75.6	−350.7	− 83.7
$SiC+3H_2O=SiO_2+CO+3H_2$	−221.2	− 52.8	−280.3	− 66.9	−327.6	− 78.2	−375.4	− 89.6
$SiC+H_2O=Si+CO+H_2$	+142.9	+ 34.1	+ 34.5	+ 8.2	− 38.5	− 9.2	−121.9	− 29.1
$SiC+H_2O=SiO+C+H_2$	+161.3	+ 38.5	+ 57.0	+ 13.6	− 13.0	− 3.1	− 77.5	− 18.5
$SiC+{}^1/_2SiO_2={}^3/_2Si+CO$	+324.7	+ 77.5	+191.5	+ 45.7	+102.2	+ 24.4	+ 5.0	+ 1.2
$SiC+SiO_2=SiO+Si+CO$	+621.4	+148.3	+364.5	+ 87.0	+192.3	+ 45.9	+ 18.8	+ 4.5
$SiC+SiO_2=2SiO+C$	+639.4	+152.7	+387.1	+ 92.4	+217.9	+ 52.0	+ 63.3	+ 15.1
$SiO+C=CO+Si$	− 18.4	− 4.4	− 22.6	− 5.4	− 25.6	− 6.1	− 44.4	− 10.6

$$SiC + 2H_2O = SiO_2 + CH_4$$

proceeds vigorously.

The oxidation of SiC by oxygen begins above 1000°C. Table 55 gives thermodynamic assessments of the possibility of various reactions taking place during oxidation [728].

From the data given there it can be seen that oxygen may oxidize SiC even at room temperature; in this process the most probable reaction is that leading to the formation of SiO_2. With water vapor reactions are possible which lead to the formation of SiO_2, CO_2, and H_2, and also of SiO_2, CO, and H_2. Because the principal oxidation product is SiO_2, whose volume is greater than that of SiC, a dense film of SiO_2 forms which is protective and retards further oxidation.

If the oxidizing actions of oxygen, water vapor, and carbon monoxide are compared, the gases can be arranged in the decreasing order of oxidizing power $O_2 - H_2O - CO$ with respect to SiC. The kinetics of oxidation of SiC have been investigated [702] in the temperature range 900-1600°C by a thermographic method. The degree of oxidation was estimated from the equation

$$\left[1 - (1 - R)^{1/3}\right]^2 = \frac{2\lambda_0^2}{r_0} K't,$$

where λ_0 is the mean diameter of the diffusing molecule, r_0 is the mean radius of the specimen, K' is the rate constant of the reaction, t is the time, and

$$R = \frac{r_0^3 - r^3}{r_0^3}.$$

In the temperature range 900-1200°C, the oxidation product is amorphous silicon oxide, whose heat of formation is 83.8 kJ/mole (20 kcal/mole). At higher temperatures it is converted into the crystalline form, whose heat of formation is 69.7 kJ/mole (16.6 kcal/mole). The presence of water vapor in the oxidizing atmosphere increases the oxidation rate.

The readily volatile SiO and CO are formed on heating SiC specimens at 900°C with an oxygen pressure of 13.33 N/m^2 (10^{-1} mm Hg) [702]. Above 900°C and with a pressure of about 0.1013 MN/m^2 (1 atm), SiO_2 is formed.

Silicon carbide readily reacts with molten carbonates, sulfates, and hydroxides of the alkali metals, B_2O_3, KNO_3, Na_2O_2, $Na_2B_4O_7$, $PbCrO_4$, $KClO_3$, Na_2SiO_3, and with mixtures of $K_2Cr_2O_7$ + $PbCrO_4$ with such oxides as CaO, MgO, CuO, FeO, NiO, MnO, Cr_2O_3, PbO, and Ag_2O [702, 831].

The thermodynamic analysis of the reaction of silicon carbide with calcium and silicon oxides [90] has shown that in the presence of these oxides silicon carbide is thermodynamically unstable in the range 1730-2130°C. In the systems SiC — SiO and SiC — SiO_2 the most probable product at temperatures up to 1900°C is silicon, and at higher temperatures SiO_2. The breakdown of silicon carbide is most likely in the reaction with SiO up to 1900°C and above that temperature with SiO_2. Moreover, reaction in the systems SiC — SiO, SiC — SiO_2, and SiC — CaO is more favorable than in the system SiC — SiO_2 — CaO (Fig. 79).

Silicon carbide, in both the polycrystalline and single-crystal forms, is decomposed by copper with the formation of a copper—silicon alloy and the separation of graphite [941].

In the analysis of commercial SiC it is usual to determine the contents of total and free silicon, silicic acid, total and free carbon, and impurities, mainly iron.

Without dwelling on those methods which have been extensively described in the literature [754, 832, 833, 837-839], we may note the methods of analyzing pure silicon carbide [702, 840-842]. These are mainly spectrographic methods in which the impurity contents are determined. The spectrographic analysis of pure SiC by the evaporation method is reported in [841]. Evaporation was carried out for 3 min at 1850°C, the spectra being recorded on a GKP-28 spectrograph. The Cu, Al, Mg, Pb, Sb, Mn, and Fe contents were determined.

The following impurities were determined in the concentration range 10^{-4}–$3 \cdot 10^{-3}$% by the spectrographic method in [842]: Cu, Ca, Mg, Al, Ti, V, and Mn.

Copper and iron present as impurities in silicon carbide were determined by a polarographic method in [840]. In this method a sample of the ground and dried material was treated with a mixture of sulfuric, hydrofluoric, and nitric acids. The insoluble residue was then treated with hydrochloric acid, after which the

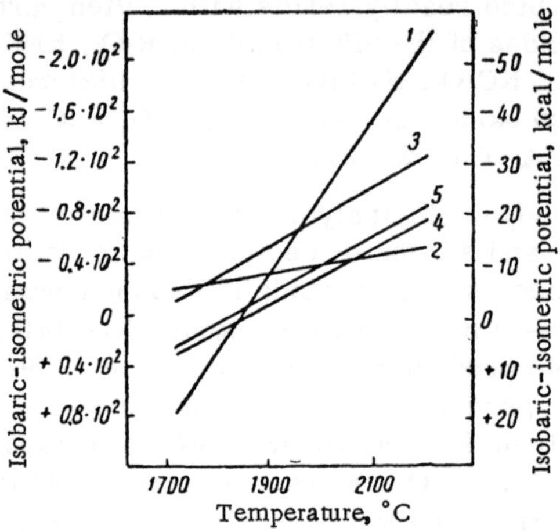

Fig. 79. Dependence of the isobaric-isothermal potential ΔZ_T^0 on the reaction temperature. 1) $SiC_{sol} + 2SiO_{liq} = SiO_{2\,gas} + CO_{gas}$; 2) $SiC_{sol} + SiO_{gas} = 2Si_{liq} + CO_{gas}$; 3) $SiC_{sol} + CaO_{sol} = CaSi_{liq} + CO_{gas}$; 4) $SiC_{sol} + \frac{1}{3}CaSiO_3 = (\frac{1}{3}Ca + \frac{4}{5}Si)_{liq} + CO_{gas}$; 5) $SiC_{sol} + \frac{1}{4}Ca_2SiO_4 = (\frac{1}{2}Ca + \frac{5}{4}Si)_{liq} + CO_{gas}$.

copper and iron were determined in aliquot parts of the filtrate. Bromphenol blue and complexone III were added to 50 ml of the solution, which was neutralized with caustic soda solution; triethanolamine and ethylenediamine were added and the solution polarographed. $E_{\frac{1}{2}}$ was 0.57 V for copper and 0.99 V for iron.

The impurities Ni, Mn, Fe, Cu, Zn, Sb, and Mo in SiC were determined by a neutron-activation method in [702]. A sample was irradiated in a reactor with a neutron flux of density $7 \cdot 10^{12}$ neutrons/cm² · sec. After irradiation the carbide sample was treated with chlorine and oxygen at 1250°C, the following reactions taking place:

$$SiC + 2Cl_2 = SiCl_4 + C,$$
$$C + O_2 = CO_2.$$

The activated impurities were separated from one another, and then their activity was measured by means of a Geiger counter.

Chapter IX

The Applications of Carbides

The properties of carbides — particularly those of the carbides of the transition metals of Mendeleev's periodic system of the elements, which are characterized by high melting points and great hardness, corrosion resistance, wear resistance, and the metallic nature of their conduction — determine the possibility of their wide application in various fields of technology. Whereas, in the early days, the transition-metal carbides were used mainly as constituents of hard alloys, within the last 10 years their fields of application have considerably broadened and now embrace, in general terms, the following:

1. Refractory materials.
2. Corrosion-resistant materials and materials for the chemical industry.
3. Materials for nuclear-energy production.
4. Materials for the manufacture of aircraft and rockets.
5. Materials for electrical and radio purposes.
6. Superhard and wear-resistant materials.

The principal fields of application of the individual carbides are given in Table 56.

TABLE 56. The Principal Fields of Application of Carbides

Materials	Phases	Principal properties	Fields of application	References
Refractories	TiC (cermet TiC−W)	Resistant to molten Na and Na−K alloys at high temperatures (1050°C) and pressures up to 861 kN/m^2 (8.5 atm)	Parts of pumps for transferring Na	[878]
	ZrC, NbC	Resistance to molten metals	Crucibles and boats for melting metals; tubes for conveying molten metals	[152]
	NbC, TaC	The same	Parts of plant for evaporating Al and Zn	
	HfC, TaC	High melting points; resistance to molten metals	In composition of special refractories for lining crucibles for melting metals	
	SiC alloys	Resistance to molten metals	Parts of plant for conveying molten Al; linings for baths for the production of Al by the electrolysis of molten salts; crucibles for melting nonferrous and rare metals; protective thermocouple sheaths	[877, 861]
Corrosion-resistant materials for the chemical industry	CaC_2	Decomposed by chemically combined water	Vigorous dehydrating agent	[116]
		Interacts with nitrogen to form calcium cyanamide	Raw material for commercial production of calcium cyanamide	[116]
	SrC_2, BaC_2, MeC_2 (Me, rare-earth metal)	Presence of double and triple bonds in compounds	Synthesis of metalloorganic compounds containing alkaline-earth and rare-earth metals	[152]
	Cr_3C_2	High resistance to various chemical reagents and resistance to oxidation	Filters in chemical industry; electrodes for use in electrochemical processes; bearings and packing in pumps supplying	[884, 886, 887]

			salt water under high pressure for washing out oil tanks of tanker vessels; parts of pumps for conveying acids; nozzles for corrosive liquids and gases	
	Cr_3C_2, Cr_7C_3, Mo_2C	Catalytic properties	Catalysts for use in organic synthesis	[889]
	SiC	High resistance to chemical reagents	Parts of pumps for transferring acid solutions; lining of nozzles for spraying corrosive liquids; condensers and scrubbers working in corrosive gas atmospheres	
Materials for nuclear-energy production	Be_2C	High melting points; strength of elevated temperatures; specific nuclear properties Addition of Be_2C to metallic Be increases its long-term strength at temperatures of 650-700°C	Constituent of structural materials; material for retarders and biological protection	[868, 869]
	UC	High melting point and thermal conductivity; absence of phase transformations; specific nuclear properties	Cores of fuel elements of nuclear reactors; fuel materials in reactors with sodium and organic heat carriers. UC–ThC alloys are used as fuel in gas-cooled reactors	[231, 242]
	ZrC	High melting point; low neutron-capture cross section	Coating for graphite matrices containing UC and ThC used as fuel elements in the power reactors for closed-cycle gas turbines	[231]

TABLE 56 (continued)

Materials	Phases	Principal properties	Fields of application	References
	NbC, TaC	High melting point; chemical resistance	Coatings for dispersion fuel elements with graphite matrices to retain fission fragments.	[231]
	B_4C	Large neutron-capture cross section	Absorbent material for regulating rods (B_4C-Al, $B_4C-Al_2O_3$, B_4C-ZrB_2, B_4C – Zircalloy)	[869]
Materials for the manufacture of aircraft and rockets	Be_2C	Addition of Be_2C to metallic Be increases its long-term strength and fracture resistance	Structural materials used in aviation	[868, 869]
	TiC	High melting point, hardness, and heat resistance; strength at high temperatures	Constituent of cermets for use in gas-turbine blades, withstanding very well the severe operating conditions experienced in aircraft jet engines; component of friction disks for aircraft construction; protective coatings for rocket components (ARD nozzles and leading parts)	[152, 872, 874, 875]
Materials for electrical and radio purposes	MeC_2 (Me, rare-earth metal)	High melting points; low work function	In composition of emitters for thermionic power transformers	[870]
	ThC	High emission current density; low work function of electrons	Cathode material that works stably at 1900°C for 900 h	[871]
	TiC	High melting point and electrical conductivity; low evaporation rate	Constituent of electrodes for the underwater oxy-electric cutting of steel; thermocouple electrodes for measuring temperatures up to 2000°C in reducing and neutral atmospheres and in vacuum	[876, 877]

ZrC	High melting point; low evaporation rate; high emission current density	Thermionic cathodes for electronic devices working in portable equipment under conditions of low vacuum and intensive ionic bombardment; constituent of cathodes (ZrC — UC) of thermionic converters of thermal energy into electrical energy	[879, 880]
NbC, TaC	High melting point; low vapor tension at high temperatures; metallic nature of conduction; satisfactory mechanical strength	Heating elements of electric resistance furnaces working in reducing and neutral atmospheres up to 3000°C	[152, 512]
TaC	High emissive power	High-intensity light source suitable for illumination for filming (1400 lm); coating on metallic tungsten and rhenium for special elements of radiation lamps	[152, 883]
Mo_2C	High melting point	Solder for high-temperature soldering in electronics	[888]
$Cr_{23}C_6$	Metallic character of conduction; small temperature coefficient of electrical resistivity	Component of standard low-ohmic resistance working at 300-400°C with nominal resistance values of 5-50 Ω	[786]
SiC	High electrical resistivity and chemical resistance; semiconducting properties	Component of firing electrodes in ignitron rectifiers; resistances intended for	[152, 709 741]

TABLE 56 (continued)

Materials	Phases	Principal properties	Fields of application	References
			protection of high-voltage transmission lines; miniature nonlinear resistors; waveguide absorbers in waveguides transmitting power at frequencies in excess of 1000 MHz; heaters for electric furnaces	
Superhard and wear-resistant materials	TiC, WC	High melting point, hardness, and heat resistance; strength at elevated temperatures; capacity for recrystallization through molten metals	Constituents of metalloceramic hard alloys ensuring higher productivity in the metal-working, mining, coal, and petroleum industries; widely used hard alloys belong to the types KA, TK, WZ, FS, and BK	[297, 567, 860, 723, 873, 890, 891, 907, 915, 916]
	ZrC, Cr_3C_2	High abrasive power	Polishing materials	[775]
	VC, NbC, TaC	High melting point, hardness, and heat resistance	Alloying addition to hard alloys based on TiC and WC; addition of VC increases tool life by 10-20%; addition of 2% TaC to alloy BK6 increases the rate of machining cast iron by 10%; addition of 4% TaC to TK-type alloys increases the rate of machining steel by 20%	[882]
	Cr_3C_2	High melting point; hardness which is retained at elevated temperatures; high chemical resistance	Constituent of filler alloys and also of hard alloys used for nozzles, dies, high-temperature bearings, press dies for forming brass sections, jets for	[873, 885-887]

			sand-spraying equipment, linings for large tube-drawing dies.	
			Coatings of carbides of Ti, V, Nb, and Cr on steel raise its wear resistance by 1.5-30 times and also considerably increase its heat and corrosion resistance.	[321]
	B_4C	Great hardness and abrasive power	Grinding and polishing hard materials; sharpening and finishing of hard-metal tool blades; cutting elements of drilling bits and tools for machining hard materials; tools for truing up grinding wheels; sand-spraying jets, gages, dies for the rod drawing of abrasive materials	[152, 691, 723]
	SiC	Great hardness and abrasive power	Grinding, polishing, and abrasive parts	[152, 723]

References

1. A Concise Chemical Encyclopedia, Vol. III, Izd. "Sovetskaya Entsiklopediya" (1964), p. 424.
2. G. V. Samsonov, Poroshkovaya Met., No. 1, p. 98 (1965).
3. G. V. Samsonov, in: Rare Alkali Metals, Izd. Sibirsk. Otd. Akad. Nauk SSSR, Novosibirsk (1965).
4. P. I. Fedorov and Su Mêng-Tsêng, Acta Chim. Sinica, 23(1): 30 (1957).
5. A. F. Alabyshev et al., Sodium and Potassium, Goskhimizdat (1959).
6. G. V. Samsonov, Refractory Compounds of Rare-Earth Metals with Nonmetals, Izd. "Metallurgiya" (1964).
7. P. Spedding et al., J. Am. Chem. Soc., 80(17): 3399 (1958).
8. T. Ya. Kosolapova et al., Zh. Neorgan. Khim., 10(11): 243 (1965).
9. G. Herzberg, Atomic Reactors and the Structure of the Atom [Russian translation], IL (1948).
10. T. J. Salzano and S. Aronson, J. Inorg. Nucl. Chem., 26: 1456 (1964).
11. G. V. Samsonov and Ya. S. Umanskii, Hard Compounds of Refractory Metals, Metallurgizdat (1957).
12. G. V. Samsonov, Izv. Sektora Fiz.-Khim. Analiza, Akad. Nauk SSSR, 27: 97 (1956).
13. A. Ya. Kuchma and G. V. Samsonov, Izv. Akad. Nauk SSSR, Neorgan. Mat., 2(11): 1970 (1966).
14. G. V. Samsonov, Ukr. Khim. Zh., 21(12): 1233 (1965).
15. G. V. Samsonov, Zh. Tekh. Fiz., 26(4): 716 (1956).
16. S. N. L'vov et al., Poroshkovaya Met., No. 4, p. 3 (1962).
17. G. V. Samsonov, Ukr. Khim. Zh., 31(10): 1006 (1965).
18. V. S. Fomenko, Emission Properties of the Elements and Chemical Compounds, Izv. "Naukova Dumka," Kiev (1964).
19. S. N. L'vov et al., Dokl. Akad. Nauk SSSR, 135(3): 577 (1960).
20. G. N. Makarenko et al., Izv. Akad. Nauk SSSR, Neorgan. Mat., 2(4): 626 (1966).
21. G. A. Meerson and G. V. Samsonov, Zh. Prikl. Khim., 25(7): 744 (1952).
22. P. Schwarzkopf and R. Kieffer, Refractory Hard Metals, Macmillan, New York (1953).

23. P. V. Gel'd and O. A. Esin, High-Temperature Reduction Processes, Metallurgizdat (1957).
24. O. Ruff and P. Wunsch, Z. Anorg. Chem., 85:292 (1914).
25. F. Friedrich and L. Sittig, Z. Anorg. Chem., 144:163 (1925).
26. C. Agte and K. Moers, Z. Anorg. Chem., 198:233 (1931).
27. G. Tammann and A. Zvorykin, Z. Anorg. Chem., 170:62 (1928).
28. G. Tammann, Z. Anorg. Chem., 126:119 (1923).
29. G. Tammann, Z. Anorg. Chem., 157:321 (1926).
30. W. Baukloh and R. Durrer, Arch. Eisenhuettenw., 4:455 (1930/31).
31. W. Baukloh and G. Zimmermann, Stahl. Eisen, 53:172 (1933).
32. W. Baukloh and R. Durrer, Z. Anorg. Chem., 222:189 (1935).
33. A. A. Baikov, Metallurg, No. 3, p. 5 (1926).
34. A. M. Gistling, Zh. Prikl. Khim., 24:266 (1951).
35. A. M. Gistling, Zh. Prikl. Khim., 25:277 (1952).
36. M. E. Pozin et al., Zh. Prikl. Khim., 27:376 (1954).
37. O. A. Esin and P. V. Gel'd, Usp. Khim., 18:658 (1949).
38. P. V. Gel'd et al., Zh. Prikl. Khim., 25:121 (1952).
39. P. V. Gel'd and G. P. Shveikin, Izv. Akad. Nauk SSSR, Otd. Tekh. Nauk, Met. i Toplivo, No. 1, p. 144 (1959).
40. R. A. Karasev et al., Izv. Akad. Nauk SSSR, Otd. Tekh. Nauk, Met. i Toplivo, No. 4, p. 94 (1956).
41. A. M. Samarin and A. A. Vertman, Tr. Inst. im. A. A. Baikova, Izd. Akad. Nauk SSSR, No. 1, p. 60 (1957).
42. A. Stadeler, Metallurgie, 5:260 (1908).
43. G. V. Samsonov, Ukr. Khim. Zh., 23(3):287 (1957).
44. W. J. Kroll and A. W. Schlechton, Trans. Electrochem. Soc., 93:274 (1948).
45. G. A. Meerson and Ya. M. Lipkes, Zh. Prikl. Khim., 14(3):291 (1941).
46. G. A. Meerson and Ya. M. Lipkes, Zh. Prikl. Khim., 18(4/5):251 (1945).
47. G. A. Meerson, Izv. Sektora Fiz. Khim. Analiza, Akad. Nauk SSSR, 16(1):213 (1943).
48. G. A. Meerson et al., Zavod. Lab., 19(10):243 (1953).
49. K. Becker, Z. Elektrochem., 34:640 (1928).
50. I. E. Campbell et al., J. Electrochem. Soc., 96:318 (1949).
51. R. Schenk et al., Z. Anorg. Chem., 203:159 (1931).
52. A. E. van Arkel, Physica, 4:286 (1924).
53. K. Moers, Z. Anorg. Chem., 198:243 (1931).
54. A. E. van Arkel and J. de Boer, Z. Anorg. Chem., 148:324 (1925).
55. P. McKenna, Metals Progr., 36:152 (1939).
56. Yu. A. Klyachko and M. M. Shapiro, Zavod. Lab., 16(10):1173 (1950).
57. N. M. Popova, Carbide Analysis of Steel, Oborongiz (1957).
58. E. S. Medvedeva, Zavod. Lab., 13(12):1413 (1947).
59. M. M. Shapiro, Zavod. Lab., 13(12):1426 (1947).
60. C. E. Wicks and F. E. Block, The Thermodynamic Properties of 65 Elements, Their Oxides, Halogenides, Carbides, and Nitrides [Russian translation], Izd. "Metallurgiya" (1965).
61. N. I. Blok et al., Zavod. Lab., 22(1):35 (1956).

62. N. I. Blok et al., Zavod. Lab., 21(8): 894 (1955).
63. L. Andrieux and G. Weiss, Bull. Soc. Chim. France, 15: 598 (1948).
64. O. Hönigschmid, Karbide und Silicide, Halle (1914).
65. U. D. Veryatin et al., The Thermodynamic Properties of Inorganic Substances, Atomizdat (1965).
66. M. P. Slavinskii, The Physicochemical Properties of the Elements, Metallurgizdat (1952).
67. K. Fredenhagen and G. Cadenbach, Z. Anorg. Chem., 158: 249 (1926).
68. H. Matignon, Compt. Rend., 124: 1026 (1897).
69. H. Moissan, Compt. Rend., 127: 911 (1898).
70. Gmelins Handbuch der anorganischen Chemie, No. 21 (1928).
71. U. Hoffman, Naturwiss., 18: 229 (1939).
72. H. Föppl, Z. Angew. Chem., 70(13): 401 (1952).
73. G. Tammann and A. Zvorykin, Z. Anorg. Chem., 168: 218 (1927).
74. K. Fredenhagen and H. Such, Z. Anorg. Chem., 178: 353 (1929).
75. H. Moissan, Compt. Rend., 126: 303 (1898).
76. H. Moissan, Compt. Rend., 136: 1221 (1903).
77. M. B. Beber and C. F. Floe, Trans. AIME, 166: 128 (1946).
78. O. Ruff and B. Bergdahl, Z. Anorg. Chem., 106: 79 (1919).
79. R. Nast and W. Pfab, Chem. Ber., 89: 415 (1956).
80. R. Nast and W. Pfab, Z, Anorg. Chem., 292: 287 (1957).
81. O. Ruff, Z. Anorg. Chem., 99: 73 (1917).
82. G. Brauer (ed.), Handbook of Preparative Inorganic Chemistry [Russian translation], IL (1956).
83. H. Nichimura and H. Kimura, J. Japan. Inst. Metals, 20: 528 (1956).
84. H. A. Sloman, J. Inst. Metals, 49: 370 (1932).
85. M. Hansen and K. Anderko, The Constitution of Binary Alloys, McGraw-Hill, New York (1958).
86. A. E. Vol, The Structure and Properties of Binary Metallic Systems, Fizmatgiz (1959).
87. M. W. Mallett et al., J. Electrochem. Soc., 101(6): 298 (1954).
88. J. H. Coobs and W. J. Koshuba, J. Electrochem. Soc., 99(3): 115 (1952).
89. M. Stackelberg and F. Quatrum, J. Phys. Chem., 27: 50 (1934).
90. I. V. Ryabchikov et al., Zh. Prikl. Khim., 37(9): 2050 (1964).
91. F. Sh. Muratov and A. V. Novoselova, Dokl. Akad. Nauk SSSR, 129(2): 334 (1959).
92. B. D. Pollock, J. Phys. Chem., 63(4): 587 (1959).
93. M. Stackelberg, Z. Elektrochem., 37: 542 (1936).
94. J. J. Neely et al., J. Am. Ceram. Soc., 33: 363 (1950).
95. M. Humenik and W. Kingery, J. Am. Ceram. Soc., 37: 18 (1954).
96. F. Sh. Muratov and A. V. Novoselova, Zh. Neorgan. Khim., 6: 1974 (1961).
97. E. Staritzky, Analyt. Chem., 28: 915 (1956).
98. J. Murakami et al., J. Japan Inst. Metals, 21: 712 (1957).
99. J. Novak, Z. Physikal. Chem., 73: 513 (1910).
100. W. H. Rueggeberg, J. Am. Chem. Soc., 65: 602 (1943).
101. A. Schneider and J. F. Cordes, Z. Anorg. Chem., 279: 94 (1955).
102. Yu. V. Baimakov and P. A. Prishchepchik, Tr. Leningrad. Ind. Inst., Issue 1, No. 4 (1940).

103. O. Kubaschewski and E. Evans, Metallurgical Thermochemistry, Pergamon Press, London (1958).
104. O. Ruff and T. Foerster, Z. Anorg. Chem., 131:321 (1923).
105. A. Ya. Rozchenko et al., Sakharn. Prom., 30(5):49 (1956).
106. M. A. Vapnik, Khim. Prom., No. 6, p. 43 (1961).
107. V. A. Ershov and I. A. Pagnueva, Zh. Prikl. Khim., 33:1901 (1961)
108. A. A. Kuznetsov and I. D. Shvarts, General Technology of Calcium Carbide and Cyanamide, ONTI (1935).
109. F. I. Egorov, Khim. Stroitel'stvo, 8:411 (1934).
110. A. A. Kuznetsov, The Production of Calcium Carbide, Goskhimizdat (1954).
111. C. Bingham, The Production of Calcium Carbide [Russian translation], Goskhimizdat (1934).
112. N. Vannerberg, Acta Chem. Scand., 16:1212 (1962).
113. M. Rothmund, Z. Anorg. Chem., 31:130 (1902).
114. H. Rudolfi, Z. Anorg. Chem., 54:170 (1907).
115. E. Schlumberger, Z. Angew. Chem., 40B:141 (1927).
116. N. A. Doronin, The Metallurgy of Calcium, Atomizdat (1959).
117. H. H. Frank et al., Z. Anorg. Chem., 232:61 (1937).
118. M. A. Bredig, Z. Physikal. Chem., 46:801 (1942).
119. M. Stackelberg, Z. Physikal. Chem., 9:437 (1930).
120. M. A. Bredig, Z. Anorg. Chem., 298:255 (1959).
121. M. A. Bredig, Z. Anorg. Chem., 310:338 (1961).
122. W. Borchert and M. Röder, Z. Anorg. Chem., 302:213 (1959).
123. P. V. Gel'd and A. G. Kologreeva, Zh. Prikl. Khim., 21:600 (1948).
124. E. Schlumberger, Z. Angew. Chem., 39B:219 (1926).
125. H. Tagawa and K. Fryimori, J. Chem. Soc. Japan, 63:1154 (1960).
126. O. Ruff and B. Josephy, Z. Anorg. Chem., 153:17 (1926).
127. J. Hockspiel and N. Platzer, Compt. Rend., 246:2969 (1958).
128. R. J. L. Andon et al., Trans. Faraday Soc., 60:1030 (1964).
129. A. N. Kornilov et al., Zh. Fiz. Khim., 38:702 (1964).
130. I. M. Fedorchenko, Poroshkovaya Met., No. 2, p. 70 (1961).
131. Yu. M. Shashkov and V. V. Rozhdestvenskaya, in: Growth of Crystals, Vol. 4, Consultants Bureau, New York (1966), p. 94.
132. S. I. Alyamovskii et al., Zh. Neorgan. Khim., 8:2000 (1963).
133. B. H. Eckstein and R. Forman, J. Appl. Phys., 33:82 (1962).
134. P. K. Gantzel and N. L. Baldwin, Acta Cryst., 17:772 (1964).
135. V. I. Alekseev and L. A. Shvartsman, Izv. Akad. Nauk Ukr SSR, Metallurgiya i Gornoe Delo, No. 2, p. 180 (1964).
136. L. B. Dubrovskaya et al., Fiz. Metallov i Metalloved., 17:73 (1964).
137. M. V. Sazonova and I. B. Ban'kovskaya, Zh. Prikl. Khim., 37:773 (1964).
138. A. S. Mikulinskii and F. S. Maron, Zh. Prikl. Khim., 33:835 (1960).
139. P. V. Gel'd and O. A. Esin, Dokl. Akad. Nauk SSSR, 63:541 (1950).
140. E. Bonnier and P. Poyet, Compt. Rend., 248:1048 (1959).
141. L. Ya. Markovskii and N. V. Vekshina, Zh. Prikl. Khim., 34:242 (1961).
142. N. V. Vekshina and L. Ya. Markovskii, Zh. Prikl. Khim., 35:30 (1962).
143. G. A. Meerson et al., Izv. Akad. Nauk SSSR, Metallurgiya i Gornoe Delo, No. 1, p. 69 (1963).

144. A. Schneider and G. Hutt, Z. Anorg. Chem., 257:289 (1949).
145. G. A. Meerson et al., At. Energ., 14: 563 (1963).
146. Gmelins Handbuch der Anorganischen Chemie, No. 30, p. 299 (1932).
147. M. Thomson, Trans. Electrochem. Soc., 16:197 (1909).
148. G. V. Samsonov et al., Zh. Obshch. Khim., 32:2753 (1962).
149. N. A. Hiel and O. B. Cavin, J. Am. Ceram. Soc., 47:360 (1964).
150. M. Hoch, J. Appl. Phys., 29:15 (1958).
151. M. De Kay-Thomson, Trans. Am. Electrochem. Soc., 54:891 (1928).
152. G. V. Samsonov, The Properties of Refractory Compounds, Metallurgizdat (1962).
153. O. Ruff and F. Keilig, Z. Anorg. Chem., 88:410 (1914).
154. Ya. I. Gerasimov and A. I. Krestovinkov, Izv. Vuz. Tsvetn. Met., No. 3, p. 54 (1958).
155. Z. Horvath, Nehezipari Muszaki Egyet. Kozl., 21:277 (1960).
156. O. A. Esin and P. V. Gel'd, Zh. Prikl. Khim., 31:986 (1958).
157. G. I. Chufarov et al., Tr. Inst. Met. Akad. Nauk SSSR, Ural'skii Filial, No. 3, p. 57 (1959).
158. G. G. Gvelesiani et al., Izv. Akad. Nauk SSSR, Otd. Tekh. Nauk, No. 8, p. 19 (1958).
159. W. Jeitschko et al., Monatsh. Chem., 95:156 (1964).
160. H. Myers, Canad. J. Phys., 35:819 (1957).
161. M. P. Slavinskii et al., Metallurg, No. 9, p. 12 (1934).
162. J. Murakami et al., J. Japan. Inst. Metals, 21:665 (1957).
163. J. H. Cox and L. M. Pidgeon, Canad. J. Chem., 41:1414 (1963).
164. O. Ruff and E. Jellinek, Z. Anorg. Chem., 97:312 (1916).
165. P. Askenasy and A. Lebedeff, Z. Electrochem., 16:559 (1910).
166. A. I. Belyaev et al., Tsvetn. Metal., 3:45 (1957).
167. Yu. V. Baimakov et al., Legkie Metal., No. 1, p. 22 (1937).
168. J. Murakami et al., New Fac. Engng. Kyoto Univ., 19:302 (1957).
169. L. Wöhler and K. Hofer, Z. Anorg. Chem., 213:249 (1933).
170. W. A. Chupka et al., J. Phys. Chem., 62:611 (1958).
171. D. Meschi and A. Searcy, J. Phys. Chem., 63:1175 (1959).
172. A. Ya. Kuchma, Izv. Akad. Nauk SSSR, Ser. Neorgan. Mat., 3:884 (1967).
173. F. D. Richardson, J. Iron Steel Inst., 175:33 (1953).
174. C. S. Campbell, Metallurg. Soc. Conf., 7:412 (1961).
175. K. Grjotheim et al., Z. Anorg. Chem., 328:267 (1964).
176. W. Jeitschko et al., Monatsh. Chem., 95:431, 436 (1964).
177. W. Jeitschko et al., J. Less-Common Metals, 7:133 (1964).
178. D. D. Jackson et al., Proc. Symp. Thermodynamics of Nuclear Materials, Vienna (1962), p. 519.
179. B. Jacobson and A. Westgren, Z. Physikal. Chem., 20:361 (1933).
180. R. C. Vickery et al., J. Chem. Soc. (1959), p. 503.
181. H. Auer-Welsbach and H. Nowotny, Monatsh. Chem., 92:198 (1961).
182. H. Nowotny and H. Auer-Welsbach, Monatsh. Chem., 92:789 (1961).
183. G. V. Samsonov et al., Dokl. Akad. Nauk SSSR, 144:1062 (1962).
184. G. V. Samsonov et al., Collection: Rare-Earth Elements, Izd. Akad. Nauk SSSR (1963).

185. G. V. Samsonov et al., Zh. Neorgan. Khim., 8: 975 (1962).
186. G. V. Samsonov et al., Zh. Prikl. Khim., 34: 1444 (1961).
187. O. Petterson, Ber., 28: 2419 (1895).
188. H. Moissan and P. Etard, Compt. Rend., 122: 543 (1897).
189. M. G. Mauzone and J. Z. Brigge, The Less-Common Alloys of Molybdenum, Climax Molybdenum Company (1962).
190. M. S. Koval'chenko et al., Zh. Prikl. Khim., 31: 1427 (1958).
191. G. N. Makarenko, Zh. Prikl. Khim., 36: 1860 (1963).
192. T. Ya. Kosolapova and G. N. Makarenko, Ukrainsk. Khim. Zh., 30: 784 (1964).
193. L. Brever and O. Krikorian, J. Electrochem. Soc., 103: 38 (1956).
194. H. Moissan, Compt. Rend., 131: 595 (1900).
195. H. Moissan, Compt. Rend., 123: 148 (1896).
196. T. Ya. Kosolapova et al., Zh. Prikl. Khim., 39: 2396 (1966).
197. H. A. Wilhelm and P. Chiotti, Trans. Am. Soc. Metals, 42: 1295 (1950).
198. H. A. Wilhelm et al., J. Chem. Soc., Suppl. Issue No. 2, p. 318 (1949).
199. Pulvermetallurgie in der Atomkerntechnik, Metallwerk Plansee AG, Reutte/Tirol (1962).
200. Tech. Conf. on Compounds of Interest in Nuclear Reactors, Colorado (1964).
201. D. Peterson, Trans. Am. Soc. Metals, 53: 765 (1961).
202. R. Mickelson and D. Peterson, Trans. Am. Soc. Metals, 50: 340 (1958).
203. L. Troost, Compt. Rend., 116: 1229 (1893).
204. H. Moissan and F. Etard, Compt. Rend., 122: 573 (1896).
205. H. Nowotny et al., Planseeber. Pulvermet., 5(3): 102 (1957).
206. D. E. Scaife and A. W. Willie, Austral. Atomic Energy Symp., Sydney (1958).
207. G. V. Samsonov et al., Zh. Prikl. Khim., 33: 1661 (1960).
208. R. S. Street and T. N. Waters, U. K. Atomic Energy Authority Rep. No. M1114 (1962).
209. C. P. Kempter and N. H. Krikorian, J. Less-Common Metals, 4: 244 (1961).
210. R. S. Street and T. N. Waters, U. K. Atomic Energy Authority Rep. No. M1115 (1962).
211. Proc. Symp. Thermodynamics of Nuclear Materials, Vienna (1962), p. 581.
212. A. Perret and J. Reithman, Helv. Chim. Acta, 26: 740 (1943).
213. P. Costa and R. Lallement, Phys. Letters, 1: 21 (1963).
214. W. Shubb, Reactor Mat., 6(2): 42 (1963).
215. E. Hund and R. Rundle, J. Am. Chem. Soc., 73: 4777 (1951).
216. H. K. Ionsdall and J. N. Graves, Proc. Symp. Thermodynamics of Nuclear Materials, Vienna (1962), p. 601.
217. H. Moissan, Compt. Rend., 122: 274 (1896).
218. P. Lebeau, Compt. Rend., 152: 755 (1911).
219. L. Litz et al., J. Am. Chem. Soc., 70: 718 (1948).
220. J. Katz and E. Rabinovich, The Chemistry of Uranium [Russian translation], IL (1954).
221. W. Wilson, J. Am. Ceram. Soc., 45: 77 (1960).
222. W. Chubb and W. M. Phillips, Trans. Am. Soc. Metals, 53: 465 (1961).
223. A. Tetsuzo et al., Sci. Papers Inst. Phys. Chem. Res., 55: 217 (1961).
224. A. Secrestet, Preparation and Properties of Uranium Monocarbide Castings, BML-1309 (1959).

225. P. Esch and A. Schneider, Z. Anorg. Chem., 257:254 (1948).
226. J. de Maria et al., J. Chem. Phys., 43:4449 (1965).
227. M. Mallett et al., J. Electrochem. Soc., 99:197 (1952).
228. J. Katz and G. T. Seaborg, The Chemistry of the Actinide Elements, Wiley, New York (1957).
229. B. G. Ignat'ev et al., At. Energ., 20:489 (1966).
230. R. W. Fischer, U. S. Pat. No. 2,580,349 (1951).
231. A. Strasser, Nuclear Eng., 5:353 (1960).
232. O. Ruff and A. Heinzelmann, Z. Anorg. Chem., 72:72 (1911).
233. O. Heusel, Z. Anorg. Chem., 154:366 (1926).
234. J. Piarra and M. B. Sinnott, J. Chem. Eng. Data, 1:451 (1962).
235. G. A. Meerson et al., At. Energ., 9(11):387 (1960).
236. O. L. Kruger, J. Am. Ceram. Soc., 46:80 (1963).
237. P. Himmelstein et al., Ber. Deutsch. Keram. Ges., 40:153 (1963).
238. New Nuclear Materials Including Nonmetallurgical Fuels, Vol. 1, Vienna (1963), p. 377.
239. B. Craven and E. J. McCarthey, J. Am. Ceram. Soc., 44:12 (1961).
240. M. P. Arbuzov and B. V. Khaenko, Poroshkovaya Met., No. 3, p. 64 (1966).
241. R. Juza and V. Wehle, Naturwiss., 52:537 (1965).
242. R. Kieffer et al., Planseeber. Pulvermet., 5(1):33 (1957).
243. A. Accary and R. Caillat, Powder Metallurgy, New York (1961), p. 209.
244. A. S. Bolgar et al., Poroshkovaya Met., No. 2, p. 100 (1966).
245. M. P. Arbuzov and B. V. Khaenko, Poroshkovaya Met., No. 4, p. 74 (1966).
246. H. Moissan, Compt. Rend., 122:1297 (1896).
247. R. Chang and C. G. Rhodes, J. Am. Ceram. Soc., 45:379 (1962).
248. Commissariat de l'Energie Atomique, Fr. 1, 193,090 (1959).
249. R. F. Stopps and J. V. Hamme, J. Am. Ceram. Soc., 47:59 (1964).
250. A. E. Austin, Acta Cryst., 12:159 (1959).
251. Tech. Rep. Internat. Atomic Energy Agency, No. 14, p. 44 (1963).
252. G. A. Meerson, At. Energ., 21:67 (1966).
253. L. N. Grossman, J. Am. Chem. Soc., 46:264 (1963).
254. L. B. Griffiths, J. Nucl. Mat., 4:336 (1961).
255. J. Norreys and M. Wheeler, Trans. Brit. Ceram. Soc., 62:180 (1963).
256. R. Mender-Penalosa and R. Taylor, J. Am. Ceram. Soc., 47:101 (1964).
257. J. Barta et al., J. Nucl. Mat., 4:222 (1961).
258. L. Toth et al., Monatsh. Chem., 92:794 (1961).
259. N. Hägg, Z. Physikal. Chem., 12:42 (1931).
260. J. M. Leitnaker and W. G. Witteman, J. Am. Phys., 36:1445 (1962).
261. L. S. Levinson, J. Chem. Phys., 38:2105 (1963).
262. H. Föppl, Z. Anorg. Chem., 291:12 (1957).
263. E. Parthe and V. Sadagopan, Monatsh. Chem., 93:263 (1962).
264. R. W. Pidd et al., J. Appl. Phys., 30:1575 (1959).
265. W. Padure, Reactor Materials, 5(4):24 (1962).
266. E. M. Prshedromirskaya and V. M. Sleptsov, Poroshkovaya Met., No. 4, p. 11 (1966).
267. M. S. Koval'chenko and V. V. Ogorodnikov, Poroshkovaya Met., No. 10, p. 48 (1966).

268. R. C. Lied and G. D. White, J. Am. Ceram. Soc., 45:149 (1962).
269. W. Zachariasen, Acta Cryst., 5:17 (1952).
270. M. A. Bredig, J. Am. Ceram. Soc., 43:493 (1960).
271. R. Chang, Acta Cryst., 14:1097 (1961).
272. J. Crauls et al., Trans. Am. Soc. Metals, 56:176 (1963).
273. G. Geisler and W. Bücher, Z. Anorg. Chem., 343:286 (1966).
274. USAEC Rep. ANI-6573 (1962); Reactor Materials, 6:28 (1963).
275. G. Wehsch and P. Prince, Reactor Materials, 6:25 (1963).
276. A. Accary and P. Blum, Nucl. Power, 5(50):122 (1960).
277. V. N. Eremenko, Titanium and Its Alloys, Izd. Akad. Nauk UkrSSR, Kiev (1960).
278. R. Jaffee et al., J. Metals, 2:1261 (1950).
279. G. V. Samsonov and V. P. Latysheva, Fiz. Metallov i Metalloved., 2:309 (1956).
280. G. V. Samsonov and V. P. Latysheva, Dokl. Akad. Nauk SSSR, 109:582 (1956).
281. G. V. Samsonov and A. P. Épik, Fiz. Metallov i Metalloved., 14:479 (1962).
282. B. Jacobson and A. Westgren, Z. Physikal. Chem., 20:362 (1933).
283. P. McKenna, Ind. Eng. Chem., 28:767 (1936).
284. Ya. S. Umanskii and S. S. Khidekel', Zh. Fiz. Khim., 15:983 (1941).
285. G. A. Meerson, Redkie Metally i Splavy, No. 4, p. 4 (1935).
286. P. Ehrlich, Z. Anorg. Chem., 259:1 (1949).
287. C. Agte and K. Moers, Z. Anorg. Chem., 198:233 (1931).
288. H. Cadoff and P. Nielsen, J. Metals, 5:248 (1953).
289. K. I. Portnoi et al., Izv. Akad. Nauk SSSR, Otd. Tekh. Nauk, Met. i Toplivo, No. 2, p. 147 (1961).
290. R. Bickerdike and G. Hugness, J. Less-Common Metals, 1:42 (1959).
291. G. Roman and G. N. Ramachandran, Current Sci., 31:321 (1962).
292. H. Moissan, Compt. Rend., 120:290 (1895).
293. O. Ruff and H. Bantzinger, Z. Anorg. Chem., 129:267 (1923).
294. C. Krainer and R. Konopicky, Berg-Hüttenmänn. Monatsh., 92:166 (1947).
295. L. Brantley and A. Beckmann, J. Am. Chem. Soc., 52:3956 (1930).
296. C. Laughlin and A. Wainer, Powder Met. Bull., 5:24 (1950).
297. J. T. Norton, Machine Design, 28:143 (1950).
298. G. Hüttig et al., Powder Met., 5:30 (1950).
299. A. Münster and W. Ruppert, Z. Elektrochem., 57:558 (1953).
300. Zh. D. Vert and M. V. Kamentsev, Zh. Neorgan. Khim., 4:17 (1959).
301. B. R. Miccioli and P. T. S. Shaffer, J. Am. Ceram. Soc., 47:351 (1964).
302. G. A. Meerson and Ya. M. Lipkes, Zh. Prikl. Khim., 12:1759 (1939).
303. V. S. Kutsev et al., Dokl. Akad. Nauk SSSR, 104:567 (1955).
304. V. I. Zhelankin et al., Zh. Neorgan. Khim., 3:1237 (1958).
305. W. Jeitschko et al., Monatsh. Chem., 95:1004 (1964).
306. H. J. Goldschmidt and J. A. Brand, J. Less-Common Metals, 5:181 (1963).
307. G. A. Meerson and O. E. Krein, Zh. Prikl. Khim., 25:134 (1952).
308. G. A. Meerson and O. E. Krein, in: The Metallurgy of Nonferrous Metals (MITsMiZ), No. 22, Metallurgizdat (1952).
309. G. V. Samsonov, Zh. Prikl. Khim., 28:1018 (1955).
310. G. V. Samsonov, Ukr. Khim. Zh., 23:287 (1957).
311. V. S. Kutsev and B. F. Ormont, Zh. Fiz. Khim., 29:597 (1955).

312. G. A. Meerson, Izv. Akad. Nauk SSSR, Otd. Tekh. Nauk, Met. i Toplivo, No. 3, p. 33 (1962).
313. D. D. Wagman et al., J. Res. Nat. Bur. Std., 34: 143 (1945).
314. A. van Arkel, Physica, 3: 76 (1923).
315. W. Burgers and J. Basart, Z. Anorg. Chem., 216: 209 (1936).
316. V. I. Alekseev and L. A. Shvartsman, Fiz. Metallov i Metalloved., 11: 545 (1961).
317. F. Pollard and P. Woodward, Trans. Faraday Soc., 46: 190 (1950).
318. V. P. Elyutin et al., Izv. Vuz. Chernaya Met., No. 11, p. 5 (1963).
319. V. P. Elyutin et al., Izv. Vuz. Chernaya Met., No. 3, p. 124 (1964).
320. V. Fattinger, The Physics of Powder Metallurgy (1951), p. 295.
321. G. V. Samsonov and A. M. Épik, Refractory-Compound Coatings, Izd. "Metallurgiya" (1964).
322. D. Schuler, Dissertation, Techn. Hochsch., Zürich (1952).
323. H. Ott, Z. Krist., 63: 1 (1926).
324. D. Moskowitz and M. Humenik, Planseeber. Pulvermet., 9: 60 (1961).
325. G. A. Meerson et al., Izv. Akad. Nauk SSSR, Otd. Tekh. Nauk, Met. i Toplivo, No. 6, p. 52 (1961).
326. Ya. S. Umanskii et al., The Physical Bases of Metal Science, GONTI (1949).
327. A. Munster, Z. Angew. Chem., 69: 281 (1957).
328. L. Brantley, Z. Krist, 77: 505 (1931).
329. H. Nowotny and R. Kieffer, Metallforsch., 2: 257 (1947).
330. W. Hoffman and A. Schrader, Arch. Eisenhüttenwes., 10: 65 (1936/37).
331. W. Dawihl and W. Rix, Z. Anorg. Chem., 244: 191 (1940).
332. E. Rudy et al., Monatsh. Chem., 91: 1761 (1960).
333. A. E. Koval'skii and T. G. Makarenko, Zh. Tekh. Fiz., 23: 265 (1953).
334. W. Rostoker, J. Metals, 4: 981 (1952).
335. R. O. Elliott and C. P. Kempter, J. Phys. Chem., 62: 630 (1958).
336. W. Chupka et al., J. Phys. Chem., 62: 5 (1958).
337. V. V. Fesenko and A. S. Bolgar, The Evaporation of Refractory Compounds, Izd. "Metallurgiya" (1965).
338. G. Humphrey, J. Am. Chem. Soc., 73: 2261 (1961).
339. S. Fujishiro, J. Japan. Soc. Powder Met., 8: 73 (1961).
340. B. Naylor, J. Am. Chem. Soc., 68: 370 (1946).
341. K. Kelley, US Bur. Mines Bull., No. 476 (1949).
342. M. T. Morozova et al., Zh. Obshch. Khim., 32: 2072 (1962).
343. R. E. Taylor, J. Am. Ceram. Soc., 44: 525 (1961).
344. M. Hoch and J. Vardi, J. Am. Ceram. Soc., 46: 245 (1963).
345. F. Glaser and D. Moskowitz, Powder Met. Bull., 6: 178 (1953).
346. N. V. Kolomoets et al., Zh. Tekh. Fiz., 28: 2382 (1958).
347. S. N. L'vov et al., Dokl. Akad. Nauk SSSR, 157: 408 (1964).
348. L. E. Hollander, J. Appl. Phys., 32: 996 (1961).
349. K. Tsuchido et al., J. Phys. Soc. Japan, 16: 2453 (1961).
350. F. Glaser and W. Ivanick, Trans. AIME, 194: 387 (1952).
351. L. L. Vereikina et al., Zavod. Lab., 26: 620 (1960).
352. P. W. Bridgman, Proc. Am. Acad. Sci., 66: 215 (1933).
353. V. N. Eremenko, Zh. Neorgan. Khim., 1: 131 (1956).

354. W. Kingery and F. Halden, Am. Ceram. Soc. Bull., 34: 117 (1955).
355. G. A. Meerson et al., Zh. Prikl. Khim., 13: 66 (1940).
356. H. Blumenthal and R. Silverman, Trans. AIME, 206: 977 (1956).
357. V. N. Eremenko and N. D. Lesnik, Poroshkovaya Met., No. 1, p. 43 (1961).
358. S. Raman and G. N. Ramachandran, Current Sci., 31: 321 (1962).
359. V. N. Eremenko and Yu. V. Naidich, Zh. Neorgan. Khim., 4: 2052 (1959).
360. G. V. Samsonov et al., Izv. Akad. Nauk SSSR, Otd. Tekh. Nauk, Met. i Toplivo, No. 5, p. 167 (1962).
361. V. V. Grigor'eva and N. N. Sereda, Poroshkovaya Met., No. 2, p. 48 (1961).
362. M. L. Baskin et al., Fiz. Metallov i Metalloved., 14: 422 (1962).
363. M. L. Baskin et al., Fiz. Metallov i Metalloved., 12: 860 (1961).
364. H. Heltz, J. Am. Ceram. Soc., 33: 340 (1950).
365. G. V. Samsonov et al., Izv. Akad. Nauk SSSR, Otd. Tekh. Nauk, Met. i Gornoe Delo, No. 4, p. 106 (1964).
366. H. Barr et al., Powder Met. Bull., 5: 62 (1950).
367. R. Kieffer and F. Kolbi, Powder Met. Bull., 4: 4 (1949).
368. O. Kubaschewsky and E. Evans, Metallurgical Thermochemistry, Pergamon, New York (1958).
369. E. Rudy and F. Benesowsky, Planseeber. Pulvermet., 8: 72 (1960).
370. Yu. B. Paderno and G. V. Samsonov, Dokl. Akad. Nauk SSSR, 137: 13 (1961).
371. M. Goldschmidt, J. Iron Steel Inst., 160: 345 (1948).
372. Ya. S. Umanskii, Carbides of Heavy Alloys, Metallurgizdat (1937).
373. A. E. Koval'skii and T. G. Makarenko, in: Microhardness, Izd. Akad. Nauk SSSR (1951).
374. G. V. Samsonov and N. S. Rozinova, Izv. Sektora Fiz. Khim. Analiza, Akad. Nauk SSSR, 27: 126 (1956).
375. F. Benesowsky and E. Rudy, Planseeber. Pulvermet., 8: 66 (1960).
376. H. E. Robson and P. W. Gieles, J. Phys. Chem., 68: 983 (1964).
377. H. Moissan and M. Lengfeld, Compt. Rend., 122: 651 (1896).
378. W. Kroll, Trans. Electrochem. Soc., 92: 187 (1947).
379. G. Verhaegen and G. Drowart, J. Chem. Phys., 37: 1367 (1962).
380. C. H. Prescott, J. Am. Chem. Soc., 48: 2534 (1926).
381. G. V. Samsonov and V. P. Latysheva, Fiz. Metallov i Metalloved., 2: 309 (1956).
382. G. V. Samsonov and A. P. Épik, Fiz. Metallov i Metalloved., 14: 479 (1962).
383. J. T. Norton and A. Mowry, Trans. AIME, 185: 133 (1949).
384. K. Becker and F. Ebert, Z. Physik, 31: 268 (1925).
385. A. E. Koval'skii and Ya. S. Umanskii, Zh. Fiz. Khim., 20: 769 (1946).
386. A. E. Koval'skii and L. A. Petrova, in: Microhardness, Izd. Akad. Nauk SSSR (1951).
387. N. V. Kolomoets et al., Zh. Tekh. Fiz., 28: 2382 (1958).
388. R. E. Taylor, J. Am. Ceram. Soc., 45: 353 (1962).
389. B. D. Pollack, J. Phys. Chem., 65: 751 (1961).
390. P. Cotter and J. Kohn, J. Am. Ceram. Soc., 37: 415 (1954).
391. V. S. Neshpor et al., Zh. Prikl. Khim., 37: 2375 (1964).
392. N. H. Krikorian et al., J. Electrochem. Soc., 110: 587 (1963).

393. V. L. Zhelankin and V. S. Kutsev, Zh. Fiz. Khim., 38:562 (1964).
394. G. V. Samsonov et al., Ogneupory, No. 7, p. 335 (1961).
395. W. Lidman and H. Ham, J. Am. Ceram. Soc., 35:236 (1952).
396. B. F. Ormont, Zh. Fiz. Khim., 33:1455 (1959).
397. V. S. Neshpor et al., Poroshkovaya Met., No. 2, p. 65 (1967).
398. E. V. Olougherty et al., Nature, 191:1194 (1961).
399. J. H. Norman and P. Winchel, J. Phys. Chem., 68:3802 (1964).
400. H. Nowotny et al., Planseeber. Pulvermet., 7:79 (1959).
401. R. Kieffer et al., Metall., 18:919 (1959).
402. V. I. Zhelankin et al., Zh. Fiz. Khim., 33:1988 (1959).
403. V. I. Zhelankin et al., Zh. Fiz. Khim., 35:2608 (1961).
404. G. A. Meerson and O. E. Krein, Zh. Neorgan. Khim., 5:1924 (1960).
405. G. V. Samsonov and V. N. Paderno, Zh. Prikl. Khim., 34:963 (1961).
406. C. Curtis et al., J. Am. Ceram. Soc., 37:458 (1954).
407. H. Nowotny et al., Monatsh. Chem., 90:669 (1959).
408. H. Nowotny et al., Monatsh. Chem., 89:701 (1958).
409. H. Nowotny and E. Laube, Planseeber. Pulvermet., 9:54 (1961).
410. K. I. Portnoi and Yu. V. Levinskii, Zh. Fiz. Khim., 37:2467 (1963).
411. V. N. Paderno, Izv. Akad. Nauk SSSR, Otd. Tekh. Nauk, Met. i Toplivo, No. 6, p. 176 (1962).
412. S. Grisaffe, J. Am. Ceram. Soc., 43:494 (1960).
413. V. S. Fomenko, Radiotekhn. i Élektron., 6:1406 (1961).
414. T. L. Matskevich et al., Zh. Tekh. Fiz., 32:1266 (1962).
415. A. Osawa and M. Oya, Kinzoku no Kenkyu, 5:34 (1928).
416. A. Westgren, Metallwirstschaft, 9:921 (1930).
417. G. Hägg, Z. Physikal. Chem., 12:51 (1933).
418. E. Mauerer et al., Arch. Eisenhuettenwes., 13:337 (1939/40).
419. N. Schönberg, Acta Chem. Scand., 8:624 (1954).
420. K. Powers and M. Doyle, Acta Metall., 6:643 (1958).
421. M. A. Gurevich and B. F. Ormont, Dokl. Akad. Nauk SSSR, 96:1165 (1954).
422. V. I. Zhelankin et al., Zh. Neorgan. Khim., 3:1237 (1958).
423. M. A. Gurevich and B. F. Ormont, Zh. Neorgan. Khim., 3:403 (1958).
424. S. I. Alyamovskii et al., Zh. Strukt. Khim., 2:445 (1961).
425. E. K. Stroms and R. J. McNeal, J. Phys. Chem., 66:1401 (1962).
426. N. M. Volkova et al., Izv. Akad. Nauk SSSR, Otd. Tekh. Nauk, Met. i Gornoe Delo, No. 5, p. 134 (1963).
427. M. A. Gurevich and B. F. Ormont, Zh. Neorgan. Khim., 2:1566 (1957).
428. H. Bittner and H. Goretzki, Monatsh. Chem., 93: 693 (1962).
429. S. Fujishiro and N. A. Gokcen, J. Electrochem. Soc., 109:835 (1962).
430. K. Becker, Z. Physik, 51:481 (1928).
431. O. Ruff and W. Martin, Z. Angew. Chem., 25:49 (1912).
432. K. Becker, Refractory Compounds and Their Application in Industry [Russian translation], ONTI (1934).
433. V. Dufek, Technike Spravy, 6:13 (1959).
434. G. A. Meerson and O. E. Krein, Zh. Neorg. Khim., 5:1924 (1960).
435. I. E. Campbell (ed.), High-Temperature Technology, Wiley, New York (1956).
436. Nuclear Reactors [Russian translation], Vol. 3, IL (1956), p. 43.

437. M. A. Gurevich, Zh. Neorgan. Khim., 8:2645 (1963).
438. V. I. Alekseev and L. A. Shvartsman, Dokl. Akad. Nauk SSSR, 133:1331 (1960).
439. P. McKenna, Ind. Eng. Chem., 28:767 (1936).
440. Ya. S. Umanskii, Zh. Fiz. Khim., 14:376 (1940).
441. E. Maurer et al., Arch. Eisenhuettenwes., 13: 337 (1939/40).
442. G. Brauer et al., Z. Anorg. Chem., 277:249 (1954).
443. G. Brauer and R. Lesser, Z. Metallkunde, 50:487 (1959).
444. S. I. Alyamovskii et al., in: Trudov Ural'skogo politekh. Inst. im. S. M. Kirova, No. 92, p. 125 (1959).
445. C. Kempter et al., J. Chem. Phys., 33:1873 (1960).
446. E. Stroms and H. Krikorian, J. Phys. Chem., 63:1749 (1959).
447. M. Nadler and C. Kempter, J. Phys. Chem., 64:1468 (1960).
448. E. Stroms and H. Krikorian, J. Phys. Chem., 64:1471 (1960).
449. P. Elliott, Trans. Am. Soc. Metals, 53:13 (1961).
450. F. G. Kusenko and P. V. Gel'd, Izv. Vuz. Tsvetnaya Met., No. 2, p. 43 (1961).
451. P. V. Gel'd and G. P. Shveikin, Izv. Akad. Nauk SSSR, Otd. Tekh. Nauk, Met. i Toplivo, No. 1, p. 144 (1959).
452. P. V. Gel'd, Usp. Khim., 29: 9 (1957).
453. V. N. Paderno and Yu. K. Lapshov, Poroshkovaya Met., No. 1, p. 75 (1963).
454. P. V. Gel'd, and V. D. Lyubimov, Poroshkovaya Met., No. 4, p. 77 (1963).
455. P. S. Kislyi et al., Poroshkovaya Met., No. 2, p. 38 (1962).
456. E. Stroms et al., Analyt. Chem., 32:1873 (1960).
457. P. V. Gel'd and F. G. Kusenko, Izv. Akad. Nauk SSSR, Otd. Tekh. Nauk, Met. i Toplivo, No. 2, p. 79 (1960).
458. G. F. Kusenko and P. V. Gel'd, Izv. Sibirsk. Otd. Akad. Nauk SSSR, No. 2, p. 46 (1960).
459. A. Mah and B. Boyle, J. Am. Chem. Soc., 77:6512 (1955).
460. A. I. Kornilov et al., Vestnik MGU, Ser. Khim., No. 6, p. 48 (1962).
461. A. I. Kornilov et al., Dokl. Akad. Nauk SSSR, 144:355 (1962).
462. G. V. Samsonov et al., Izv. Akad. Nauk SSSR, Otd. Tekh. Nauk, Met. i Toplivo, No. 1, p. 142 (1961).
463. R. J. Fries, J. Chem. Phys., 37:320 (1962).
464. P. V. Gel'd and V. A. Tskhai, Some Questions Regarding the Crystallochemistry of Cubic Oxides and Carbides of Titanium, Vanadium, and Niobium. Report of a Conference of the Permanent Interinstitute Colloquim on Solid Phases of Variable Composition, No. 32(2), Leningrad (1962).
465. F. E. Ellinger, Trans. Am. Soc. Metals, 31:89 (1943).
466. R. Lesser and G. Brauer, Z. Metallkunde, 49:622 (1958).
467. V. I. Smirnova and B. F. Ormont, Dokl. Akad. Nauk SSSR, 96:557 (1954).
468. V. I. Smirnova and B. F. Ormont, Dokl. Akad. Nauk SSSR, 100:127 (1955).
469. V. I. Smirnova and B. F. Ormont, Zh. Fiz. Khim., 30:1327 (1956).
470. D. Robins, Z. Agnew. Chem., 70B: 474 (1958).
471. K. K. Kelley, Trans. Am. Soc. Steel Treat., 19:233 (1932).
472. L. P. Mal'kov and A. V. Khokhlova, Redkie Metally, No. 1, p. 4 (1935).
473. P. Cooper et al., 18th Internat. Congr. Pure Appl. Chem., Montreal, Canada (1961).

474. R. Genders and R. Harrison, J. Iron Steel Inst., No. 2, p. 202 (1936).
475. M. Nacken et al., Arch. Eisenhuettenwes., 33: 635 (1962).
476. B. Lersmacher and S. Scholz, Arch. Eisenhuettenwes., 32: 421 (1961).
477. K. Becker and F. Ebert, Z. Techn. Phys., 148:216 (1930).
478. R. Born, Nature, 145: 741 (1940).
479. J. T. Norton and A. Mowry, Trans. AIME, 185:133 (1949).
480. G. V. Samsonov and V. B. Rukina, Dokl. Akad. Nauk URSR, p. 247 (1957).
481. O. P. Kolchin and I. K. Berlin, Tsvetn. Metal., 8:67 (1964).
482. I. L. Zagyanskii et al., Dokl. Akad. Nauk SSSR, 74: 723 (1950).
483. O. Ruff and T. Foehr, Z. Anorg. Chem., 104:27 (1918).
484. A. Yegesack, Z. Anorg. Chem., 154:30 (1926).
485. O. Nischk, Z. Elektrochem., 29:373 (1923).
486. O. Ruff, Z. Elektrochem., 29:459 (1923).
487. B. Kraiczek and F. Sauerwald, Z. Anorg. Chem., 185:193 (1930).
488. E. Friemann and F. Sauerwald, Z. Anorg. Chem., 203:64 (1932).
489. J. Westbrook, J. Metals, 9:1227 (1957).
490. A. Westgren and G. Phragmen, Z. Anorg. Chem., 187:401 (1930).
491. K. Hatsuta, Kinzoku no Kenkyu, 8:88 (1931).
492. D. Bloom and N. Grant, Trans. AIME, 188:41 (1951).
493. G. Tammann and R. Schönert, Z. Anorg. Chem., 122:27 (1922).
494. W. Smith, J. Metals, 9:47 (1957).
495. P. Sauerwald et al., Z. Anorg. Chem., 210:21 (1933).
496. P. V. Gel'd and O. A. Esin, Zh. Prikl. Khim., 23:1260 (1950).
497. K. Kelley et al., Tech. Paper No. 662 (1944).
498. O. Kubaschewsky and F. Wittig, Z. Elektrochem., 47:433 (1941).
499. V. P. Elyutin et al., Ferroalloys, Metallurgizdat (1951), p. 181.
500. E. Z. Vintaikin, Fiz. Metallov. i Metalloved., 16:144 (1963).
501. F. Ito, Edike Yaziko, 13: 15 (1956).
502. B. Cech, Hutnicke Listy, 13:113 (1958).
503. B. Cech, Hutnicke Listy, 13:955 (1958).
504. T. Ya. Kosolapova and G. V. Samsonov, Zh. Prikl. Khim., 32:55 (1959).
505. T. Ya. Kosolapova and G. V. Samsonov, Zh. Prikl. Khim., 32:1505 (1959).
506. G. V. Samsonov and T. Ya. Kosolapova, Powder Metallurgy in Mechanical Engineering, Izd. NTO Mashprom., Kiev (1961), p. 28.
507. T. Ya. Kosolapova and G. V. Samsonov, Zh. Prikl. Khim., 32:1704 (1960).
508. A. A. Vertman and A. M. Samarin, Transactions of a Conference on the Application of Vacuum in Metallurgy (1958), p. 132.
509. B. Owen and R. Weber, Am. Inst. Met. Eng., Tech. Paper No. 203 (1948).
510. V. I. Arkharov and V. N. Konev, Vestn. Mashinostr., 11:55 (1955).
511. V. N. Konev, Investigations on Heat-Resistant Alloys, Vol. 3 (1958), p. 415.
512. L. F. Mal'tseva et al., Poroshkovaya Met., No. 11, p. 87 (1965).
513. A. Arnold and A. Read, J. Iron Steel Inst., 73:249 (1911).
514. V. A. Tskhai and P. V. Gel'd, Fiz. Metallov i Metalloved., 16:493 (1963).
515. W. Crafts and J. Lamont, Trans. AIME, 180:561 (1950).
516. F. Wever and W. Koch, Arch. Eisenhuettenwes., 21:143 (1950).
517. J. Brown and D. Clark, Nature, 167:728 (1951).

518. S. N. L'vov et al., Fiz. Metallov i Metalloved., 11: 14 (1961).
519. P. I. Mel'nichuk et al., Fiz. Metallov i Metalloved., 9: 918 (1960).
520. A. Z. Men'shikov and P. T. Rudomanov, Tr. Inst. Fiz. Metallov, Akad. Nauk SSSR, No. 22, p. 69 (1959).
521. V. A. Tskhai and P. V. Gel'd, Zh. Strukt. Khim., 5: 275 (1964).
522. A. Westgren and G. Phragmen, Z. Anorg. Chem., 156: 27 (1926).
523. T. Takei, Sci. Rep. Tohoku Univ., No. 17, p. 939 (1928).
524. S. Hoyt, Trans. Am. Inst. Min. Met. Engrs., 89: 9 (1930).
525. W. Few and G. Manning, Trans AIME, 194: 271 (1952).
526. J. Lander and K. Garmer, Am. Inst. Min. Met. Engrs., Paper No. 2259 (1947).
527. H. Moissan and M. Hoffman, Compt. Rend., 138: 1558 (1904).
528. H. Tutuja, Sci. Papers Inst. Phys. Chem. Res. Tokyo, 19: 384 (1932).
529. Gmelins Handbuch für anorganischen Chemie, No. 53 (1935).
530. W. Dawihl, Z. Anorg. Chem., 262: 212 (1950).
531. H. Nowotny and R. Kieffer, Z. Anorg. Chem., 267: 261 (1952).
532. W. P. Sykes et al., Trans. Am. Inst. Min. Met. Engrs., 117: 175 (1935).
533. H. Nowotny et al., Monatsh. Chem., 85: 255 (1954).
534. K. Kuo and G. Hägg, Nature, 170: 245 (1952).
535. E. Rudy and F. Benesovsky, Planseeber. Pulvermet., 10: 42 (1962).
536. E. Rudy et al., Z. Metallkunde, 54: 345 (1963).
537. V. N. Eremenko and T. Ya. Velikanova, Poroshkovaya Met., No. 1, p. 41 (1965).
538. N. M. Volkova et al., Zh. Neorgan. Khim., 10: 1758 (1965).
539. O. Meyer, Arch. Eisenhuettenwes., 4: 196 (1930/31).
540. A. S. Bolgar et al., Poroshkovaya Met., No. 1, p. 40 (1967).
541. A. J. Hegedüs and J. Neugebauer, Z. Anorg. Chem., 305: 216 (1960).
542. R. F. Voitovich and N. P. Shakhanova, Poroshkovaya Met., No. 3, p. 75 (1967).
543. O. Meyer and W. Eilender, Arch. Eisenhuettenwes., 11: 545 (1938).
544. L. Browning and P. Emmett, J. Am. Chem. Soc., 74: 4773 (1952).
545. A. A. Ravdel', Zh. Russk. Fiz.-Khim. Obshchestva, 62: 1515 (1930).
546. G. Weiss, Ann. Chim., 1: 446 (1946).
547. J. Arnold and A. Read, Proc. Inst. Mech. Eng., No. 2, p. 223 (1914).
548. H. Nowotny and R. Kieffer, Metallforsch., 2: 257 (1947).
549. B. Fries and C. Kempter, Analyt. Chem., 32: 1898 (1960).
550. M. Gleiser and J. Chipman, J. Phys. Chem., 66: 1539 (1962).
551. V. I. Alekseev and L. A. Shvartsman, Izv. Akad. Nauk SSSR, Otd. Tekh. Nauk, Met. i Toplivo, No. 6, p. 171 (1962).
552. Yu. B. Paderno et al., Elektronika, No. 4, p. 165 (1959).
553. A. M. Belikov and Ya. S. Umanskii, Kristallografiya, 4: 684 (1959).
554. W. P. Sykes, Trans. Am. Soc. Steel Treat., 18: 968 (1930).
555. L. A. Tret'yachenko and V. N. Eremenko, Izv. Akad. Nauk SSSR, Ser. Neorgan. Mat., 2: 1568 (1966).
556. E. A. Voronkova, in: Hard Alloys (VNIITS), no. 1, p. 139 (1959).
557. F. Skaupy, Z. Elektrochem., 33: 487 (1927).
558. G. S. Kreimer et al., Zh. Tekh. Fiz., 22: 848 (1952).
559. A. F. Gerasimov et al., Fiz. Metallov i Metalloved., 11: 596 (1961).
560. V. V. Karpukhin et al., Tsvetn. Metal., No. 9, p. 67 (1957).

561. G. A. Meerson et al., Metallurgy of the Rare Metals, Metallurgizdat (1955).
562. C. Agte and J. Hruska, Hutnicke Listy, 9:642 (1954).
563. H. Booss, Arch. Eisenhuettenwes., 30:761 (1959).
564. A. Hegedüs and S. Gado, Z. Anorg. Chem., 305:227 (1960).
565. A. Chretian, Compt. Rend., 234:2608 (1952).
566. D. Hurd et al., Ind. Eng. Chem., 44:2432 (1952).
567. H. M. Greenhouse et al., J. Am. Ceram. Soc., 37:203 (1954).
568. Machinist, 98:375 (1954).
569. A. Newskiz and J. Aliferis, J. Am. Chem. Soc., 79:4629 (1957).
570. A. P. Komar and Yu. N. Talinin, Izv. Akad. Nauk SSSR, Ser. Fiz., 22:580 (1958).
571. P. Pfey and W. Rix, Z. Metallkunde, 45:116 (1954).
572. L. N. Butorina, Kristallografiya, 5:234 (1960).
573. V. I. Alekseev and L. A. Shvartsman, Izv. Akad. Nauk SSSR, Otd. Tekh. Nauk, Met. i Gornoe Delo, No. 1, p. 91 (1963).
574. V. I. Alekseev and L. A. Shvartsman, Dokl. Akad. Nauk SSSR, 133:1331 (1960).
575. L. Kh. Pivovarov et al., Fiz. Metallov i Metalloved., 17:606 (1964).
576. Yu. V. Zubenko, Fiz. Tverd. Tela, 3:528 (1961).
577. O. Ruff and W. Bormann, Z. Anorg. Chem., 88:365 (1914).
578. R. Vögel and W. Döring, Arch. Eisenhuettenwes., 9:247 (1935/36).
579. J. C. Southard and G. E. Moore, J. Am. Chem. Soc., 64:1769 (1942).
580. K. Kuo and E. Perssom, J. Iron Steel Inst., 178:39 (1954).
581. B. Jacobson and A. Westgren, Z. Physikal. Chem., 20:361 (1933).
582. A. Westgren, Jernkontorets Ann., 177:501 (1933).
583. R. Schenock and K. Meyer, Z. Anorg. Chem., 239:161 (1938).
584. F. N. Tavadze, Tr. Inst. Metal. i Gorn. Dela, Akad. Nauk Gruz. SSR, 1:197 (1947)
585. M. Isobe, Sci. Rep. Res. Inst. Tohoku Univ., A3:468 (1951).
586. F. N. Tavadze et al., Soobshch. Akad. Nauk Gruz.SSR, 16:477 (1955).
587. V. I. Alekseev and L. A. Shvartsman, Dokl. Akad. Nauk SSSR, 141:346 (1961).
588. M. Picon and J. Flahaut, Compt. Rend., 245:534 (1957).
589. H. Ulich and H. Siemonsen, Arch. Eisenhuettenwes., 14:27 (1940).
590. O. Ruff and E. Gersten, Chem. Ber., 46:400 (1913).
591. W. A. Roth and R. Grau, Arch. Eisenhuettenwes., 3:339 (1929/30).
592. M. Kh. Karapet'yants, Chemical Thermodynamics, Goskhimizdat (1953).
593. K. K. Kelley and G. E. Moore, J. Am. Chem. Soc., 65:782 (1943).
594. K. K. Kelley, Bull. US Bur. Mines, No. 407 (1935).
595. C. L. McCabe et al., Trans. AIME, 209:17 (1957).
596. V. I. Alekseev and L. A. Shvartsman, Dokl. Akad. Nauk SSSR, 133:1331 (1960).
597. H. Hendrus et al., Z. Metallkunde, 48:615 (1957).
598. S. Fujishiro, Trans. Japan. Inst. Metals, 1:125 (1960).
599. W. L. Worrel and J. Chipman, J. Phys. Chem., 68:860 (1964).
600. E. J. Huber et al., J. Phys. Chem., 65:1846 (1961).
601. W. Trzebiatowski, Z. Anorg. Chem., 233:327 (1937).
602. J. Hughes, J. Less-Common Metals, 1:377 (1959).
603. A. I. Evstyukhin et al., Metallurgy and Metal Science of Pure Metals, No. 4, Goskhimizdat (1963), p. 149.

604. Yu. B. Kuz'ma et al., Poroshkovaya Met., No. 4, p. 40 (1963).
605. W. Trzebiatowski and J. Rubdinski, Z. Chem., 2:158 (1962).
606. S. Epstein, The Alloys of Iron and Carbon, Vol. 1: Constitution, New York (1936).
607. K. B. Bunin, Iron–Carbon Alloys, Mashgiz, Kiev (1949).
608. W. Glund et al., Ber., 62:2483 (1929).
609. U. Hoffmann and E. Croll, Z. Anorg. Chem., 191:414 (1930).
610. L. J. E. Hofer et al., J. Am. Chem. Soc., 71:189 (1949).
611. G. Hägg, Z. Krist., A89:92 (1934).
612. Z. G. Pinsker and S. V. Kaverin, Kristallografiya, 2:386 (1957).
613. J. Moreau et al., Rev. Met., 48:486 (1951).
614. S. T. Rostovtsev et al., Stal, No. 8, p. 679 (1953).
615. G. I. Chufarov and B. D. Averbukh, Zh. Fiz. Khim., 5:1292 (1934).
616. Z. G. Pinsker and S. V. Kaverin, Kristallografiya, 1:66 (1956).
617. V. A. Apaev, Fiz. Metallov. i Metalloved., 12:409 (1961).
618. M. André, Bull. Soc. Chim. France, No. 1, p. 143 (1961).
619. S. Nagakuro, J. Phys. Soc. Japan, 14:186 (1959).
620. N. N. Sirota, Dokl. Akad. Nauk SSSR, 64:697 (1949).
621. R. Reuer and R. Kesper, Ferrum, 11:257 (1913/14).
622. F. Adcock, J. Iron Steel Inst., 135:281 (1937).
623. J. H. Andrew and D. Binnie, J. Iron Steel Inst., 119:309 (1929).
624. G. B. Harris and W. Hume-Rothery, J. Iron Steel Inst., 174:212 (1953).
625. L. S. Darken and R. W. Gurry, Trans. AIME, 191:1015 (1951).
626. R. Ruer and F. Goernes, Ferrum, 14:161 (1916/17).
627. E. Piwowarosky, Stahl Eisen, 54: 82 (1934).
628. H. Ruer and J. Berlin, Z. Anorg. Chem., 113:98 (1920).
629. G. Rümelin and R. Maire, Ferrum, 12:141 (1914/15).
630. P. Bardenheuer, Ferrum, 14:129 (1916/17).
631. F. Stäblein, Stahl Eisen, 46:101 (1926).
632. H. Esser, Stahl Eisen, 47:337 (1927).
633. R. F. Mehl and C. Wells, Trans. AIME, 125:429 (1937).
634. N. J. Wark, Stahl Eisen, 31:2108 (1911).
635. N. Tchischewsky and N. Schulgin, J. Iron Steel Inst., 95:189 (1917).
636. N. Gutovskii, Izv. Tomsk. Tekhnol. Inst., 36:1 (1914).
637. W. Köster, Z. Metallkunde, 22:289 (1930).
638. S. Tamura, J. Iron Steel Inst., 115:747 (1927).
639. N. P. Picling, Trans. Am. Inst. Min. Met. Eng., 43:145 (1924).
640. É. Z. Kaminskii and T. I. Stelletskaya, Probl. Metalloved. i Fiz. Metall., No. 2, p. 176 (1951).
641. V. N. Gridnev, Zh. Tekh. Fiz., 16:915 (1946).
642. J. K. Stanley, J. Metals, 1:752 (1949).
643. C. A. Wert, J. Metals, 2:1242 (1950).
644. E. Lindstrand, Acta Metall., 3:431 (1955).
645. E. F. Petrova et al., Dokl. Akad. Nauk SSSR, 121:1021 (1958).
646. M. P. Arbuzov and E. G. Aizentsok, Dokl. Akad. Nauk UkrSSR, No. 2, p. 150 (1955).

647. B. Migand and J. Talbot, Compt. Rend., 245:2282 (1958).
648. B. I. Bruk, Dokl. Akad. Nauk SSSR, 128:709 (1959).
649. M. P. Arbuzov, Dokl. Akad. Nauk SSSR, 74:1085 (1950).
650. B. A. Apaev, Dokl. Akad. Nauk SSSR, 93:647 (1953).
651. V. N. Gridnev, Zh. Tekh. Fiz., 24:1812 (1954).
652. V. N. Gridnev and V. I. Trefilov, Dokl. Akad. Nauk SSSR, 96:741 (1954).
653. S. S. Shteinberg, The Heat Treatment of Steel, Metallurgizdat (1945).
654. S. F. Yur'ev, Specific Volumes in the Martensitic Transformation of Austenite, Metallurgizdat (1950).
655. V. D. Sadovskii, in: Phase Transformations in Iron–Carbon Alloys, Mashgiz (1950).
656. O. Ruff, Ber., 45:69 (1912).
657. R. Schenk et al., Z. Anorg. Chem., 164:313 (1927).
658. S. M. Gutman, Zavod. Lab., No. 12, p. 1403 (1947).
659. Guderman et al., Zavod. Lab., No. 11/12, p. 900 (1946).
660. J. E. Hofer and W. C. Peebles, J. Am. Chem. Soc., 69:893 (1947).
661. N. M. Popova and M. F. Rybina, Zavod. Lab., No. 12, p. 1412 (1947).
662. J. Arnold and A. Read, J. Soc. Chem. Ind., 65:788 (1894).
663. D. Meinhardt and O. Krisemant, Arch. Eisenhuettenwes., 33:493 (1962).
664. O. Kubaschewsky and J. A. Catterall, Thermochemical Data of Alloys, Pergamon Press, London (1956).
665. S. V. Radcliffe and E. C. Rollason, J. Iron Steel Inst., 189:45 (1958).
666. G. Boecker, Metallurgie, 9:296 (1912).
667. H. Morrogh and W. J. Williams, J. Iron Steel Inst., 155:321 (1947).
668. H. Bahr and V. Jessen, Ber., 65:2226 (1930).
669. J. Drain, Ann. Chim., 8:900 (1953).
670. R. Juza and H. Puff, Naturwiss., 38:331 (1951).
671. W. Hempel, Z. Anorg. Chem., 17:296 (1904).
672. K. Friedenreich and A. Leroux, Metallurgie, 7:10 (1910).
673. S. Fujishiro and N. A. Gocken, J. Phys. Chem., 65:161 (1961).
674. J. J. Lander et al., J. Appl. Phys., 23:1305 (1952).
675. W. Baukloh and F. Springorum, Z. Anorg. Chem., 230:315 (1937).
676. J. Smidt, Z. Anorg. Chem., 216:85 (1933).
677. A. A. Vertman et al., Izv. Akad. Nauk SSSR, Otd. Tekh. Nauk, Met. i Toplivo, No. 6, p. 37 (1962).
678. A. A. Vertman et al., Dokl. Akad. Nauk SSSR, 148:342 (1963).
679. I. S. Miroshnichenko, Izv. Vuz. Tsvetn. Met., No. 1, p. 128 (1961).
680. H. A. Bahr and Th. Bahr, Ber., 61:2177 (1928).
681. B. Jacobson and A. Westgren, Z. Physikal. Chem., 20:361 (1933).
682. R. Ridgway, Trans. Am. Electrochem. Soc., 66:117 (1934).
683. B. F. Ormont, The Structures of Inorganic Substances, Gostekhizdat (1950).
684. G. A. Meerson and G. V. Samsonov, Izv. Sektora Fiz. Khim. Analiza, Akad. Nauk SSSR, 22:92 (1953).
685. G. S. Zhdanov et al., Zh. Fiz. Khim., 28:1076 (1954).
686. F. Glaser et al., J. Appl. Phys., 24:731 (1958).
687. V. M. Kudryavtsev and G. V. Samsonov, Byul. Seminara po Zharostoikim Materialam, Izd. Akad. Nauk UkrSSR, No. 6, p. 52 (1960).

688. G. V. Samsonov et al., Fiz. Metallov i Metalloved., 3:309 (1956).
689. G. V. Samsonov et al., Zh. Fiz. Khim., 32:2424 (1958).
690. N. N. Zhuravlev et al., Izv. Akad. Nauk SSSR, Otd. Tekh. Nauk, Met. i Toplivo, No. 1, p. 133 (1961).
691. G. V. Samsonov et al., Boron, Its Compounds and Alloys, Izd. Akad. Nauk UkrSSR, Kiev (1960).
692. W. Dawihl, Z. Metallkunde, 43:138 (1952).
693. G. A. Meerson and G. V. Samsonov, in: Boron (Proceedings of a Conference on the Chemistry of Boron and Its Compounds), Gostekhizdat (1958), p. 52.
694. L. Ya. Markovskii et al., Chemical Electrothermy, Goskhimizdat (1952).
695. F. K. Aleinikov, Zh. Tekh. Fiz., 27:567 (1953).
696. P. S. Kislyi and G. V. Samsonov, Fiz. Tverd. Tela, 2:692 (1960).
697. K. K. Kelley, J. Am. Chem. Soc., 63:1137 (1941).
698. R. Colson, Compt. Rend., 94:1316 (1882).
699. G. Tammann, Z. Anorg. Chem., 115:141 (1921).
700. H. Nowotny et al., Monatsh. Chem., 85:259 (1954).
701. I. S. Brokhin and V. F. Funke, Zh. Neorgan. Khim., 3:847 (1958).
702. J. R. O'Connor and J. Smiltens (eds.), Silicon Carbide, A High-Temperature Semiconductor, Pergamon Press, New York (1960).
703. A. I. Avgustinik et al., Izv. Akad. Nauk SSSR, Ser. Neorgan. Mat., 2:1439 (1966).
704. M. V. Kamentsev, Artificial Abrasive Materials, Mashgiz (1950).
705. G. M. Ippolitov and E. P. Kakushadze, The Production of Corundum and Carborundum and Their Application, ONTI, Mashmetizdat (1933).
706. F. A. Fitz-Gerald, Carborundum, Halle (1924).
707. D. L. Orshanskii, Electrothermy, GONTI, Moscow (1939).
708. H. Kiammroth, Chemische Untersuchungen an Siliziumkarbid-Heizstaben, Düsseldorf (1935).
709. G. V. Samsonov, Silicides and Their Use in Technology, Izd. Akad. Nauk UkrSSR, Kiev (1959).
710. A. S. Berezhnoi, Silicon and Its Binary Systems, Consultants Bureau, New York (1960).
711. S. A. Dobrolezh et al., Silicon Carbide, Gos. Izd. Tekhn. Lit. UkrSSR, Kiev (1963).
712. G. G. Gnesin and F. F. Lezhenin, Poroshkovaya Met., No. 2, p. 36 (1967).
713. A. N. Novikov, Zh. Prikl. Khim., 20:431 (1947).
714. A. Dietzel et al., Ber. Deut. Keram. Ges., 37:524 (1960).
715. V. S. Neshpor et al., Izv. Akad. Nauk SSSR, Ser. Neorgan. Mat., 2:855 (1966).
716. L. M. Adelshevg et al., J. Am. Ceram. Soc., 49:573 (1966).
717. H. N. Baumann, J. Electrochem. Soc., 99:109 (1952).
718. J. Chipman et al., Acta Metall., 2:439 (1954).
719. R. N. Hall, J. Appl. Phys., 29:914 (1958).
720. G. M. Zaretskaya et al., Izv. Akad. Nauk SSSR, Otd. Tekh. Nauk, Met. i Gornoe Delo, 4:58 (1964).
721. I. G. Shafran and R. R. Vladimirskaya, Zavod. Lab., 7:169 (1938).

722. I. G. Shafran, Zh. Prikl. Khim., 11: 561 (1938).
723. G. V. Samsonov and K. I. Portnoi, Alloys Based on Refractory Compounds, Oborongiz (1961).
724. T. Kendall, J. Chem. Phys., 21: 821 (1953).
725. G. A. Meerson et al., Izv. Akad. Nauk SSSR, Otd. Tekh. Nauk, Met. i Gornoe Delo, No. 1, p. 67 (1964).
726. A. Taylor and D. Laidler, Brit. J. Appl. Phys., 7: 174 (1950).
727. G. H. Fetterley, J. Electrochem. Soc., 104: 322 (1957).
728. A. Dietzel et al., Ber. Deut. Keram. Ges., 37: 524 (1960).
729. H. N. Baumann, J. Electrochem. Soc., 99: 109 (1952).
730. E. Whitney, Nature, 199: 278 (1963).
731. H. Ott, Z. Krist., 61: 515 (1925).
732. N. V. Belov, The Structure of Ionic Crystals and Metallic Phases, Izd. Akad. Nauk SSSR (1947).
733. G. S. Zhdanov and Z. V. Minervina, Zh. Éksperim. i Teor. Fiz., 17: 1 (1947).
734. G. S. Zhdanov and Z. V. Minervina, Zh. Éksperim. i Teor. Fiz., 15: 155 (1945).
735. L. S. Ramsdell and J. A. Kohn, Acta Cryst., 5: 215 (1952).
736. O. Ruff, Trans. Electrochem. Soc., 68: 87 (1935).
737. O. Ruff and M. Konchak, Z. Elektrochem., 32: 517 (1926).
738. D. V. Badami, Nature, 193: 569 (1962).
739. G. A. Meerson et al., Izv. Akad. Nauk SSSR, Ser. Neorgan. Mat., 2: 604 (1966).
740. J. Drowart and G. de Maria, Proceedings of the Conference on Silicon Carbide, Boston, Mass., Pergamon Press, Oxford (1960).
741. R. H. Rein and J. Chipman, J. Phys. Chem., 67: 839 (1963).
742. J. C. D'Entramont and J. Chipman, J. Phys. Chem., 67: 499 (1963).
743. A. Fal'kevich, Khimstroi, 6: 442 (1934).
744. H. Tagawa and K. Fryimori, J. Chem. Soc. Japan, 65: 167 (1962).
745. N. A. Gol'dberg and Yu. D. Znamenskii, Zh. Fiz. Khim., 36: 2748 (1962).
746. N. A. Gol'dberg and Yu. D. Znamenskii, Dokl. Akad. Nauk SSSR, 110: 1048 (1956).
747. N. A. Gol'dberg and Yu. D. Znamenskii, Dokl. Akad. Nauk SSSR, 120: 140 (1958).
748. E. Biesalski and H. Eck, Z. Anorg. Chem., 156: 226 (1926).
749. E. Ebler and W. Bender, Z. Anorg. Chem., 88: 255 (1914).
750. J. Rothe and F. Hinrichsen, Chem. Zentr., 78: 992 (1907).
751. A. G. Bogdanchenko, Zavod. Lab., 15: 851 (1949).
752. A. G. Bogdanchenko, Zavod. Lab., 29: 163 (1963).
753. Gmelins Handbuch der anorganischen Chemie, No. 41 (1951).
754. The Analysis of Refractory Compounds, Metallurgizdat (1962).
755. V. P. Kopylova, Zh. Prikl. Khim., 33: 1936 (1961).
756. F. Wagner et al., Trans. Am. Soc. Metals, 48: 742 (1956).
757. G. V. Samsonov and N. K. Golubeva, Zh. Fiz. Khim., 30: 258 (1956).
758. R. F. Strickland-Constable, Trans. Faraday Soc., 40: 333 (1944).
759. R. F. Strickland-Constable, Trans. Faraday Soc., 43: 769 (1947).
760. H. J. Boosz, Metall, 10: 130 (1956).
761. A. N. Selikman and N. N. Gorovits, Zh. Prikl. Khim., 25: 134 (1952).

762. A. Münster, Z. Angew. Chem., 69: 281 (1957).
763. H. J. Boosz, Metall, 7: 40 (1962).
764. A. Kh. Klibus and T. N. Nazarchuk, Zh. Anal. Khim., 16: 79 (1961).
765. S. A. Reed et al., Analyt. Chim. Acta, 10: 429 (1954).
766. J. W. Mellor, A Comprehensive Treatise on Inorganic and Theoretical Chemistry, Longmans and Green and Co., Inc., New York (1922-1937).
767. A. I. Evstyukhin et al., in: The Metallurgy and Metal Science of Pure Metals, No. 1 (1959).
768. J. Niemiec, Compt. Rend., 251: 875 (1960).
769. M. Tetenbaum et al., Trans. Am. Nucl. Soc., 4: 89 (1961).
770. P. Lebeau and A. Damiens, Compt. Rend., 156: 1987 (1913).
771. C. Kempter and M. Nadler, J. Chem. Phys., 33: 1580 (1960).
772. C. Kempter, J. Chem. Phys., 41: 1515 (1964).
773. Gmelins Handbuch für anorganischen Chemie, No. 65 (1942).
774. K. A. Peakall and J. E. Antill, J. Less-Common Metals, 4: 426 (1962).
775. A. Ya. Artamonov et al., Poroshkovaya Met., No. 2, p. 29 (1967).
776. E. W. Murbach, Trans. Met. Soc. AIME, 227: 487 (1963).
777. J. Besson et al., Compt. Rend., 258: 4079 (1964).
778. U. K. Atomic Energy Authority Production Group Rep. No. 444 (1963).
779. A. Combarien et al., Compt. Rend., 256: 5518 (1963).
780. H. D. Witzke, Abhandl. Deut. Akad. Wiss. Berlin, Kl. Math. Physik Techn., No. 6, p. 86 (1962).
781. G. E. J. Beckman, J. Electrochem. Soc., 110: 84 (1963).
782. J. P. Bouchaud and R. Fruchart, Compt. Rend., 258: 3495 (1964).
783. H. Haschke et al., Monatsh. Chem., 97: 1045 (1966).
784. L. V. Pavlinov and V. N. Bykov, Fiz. Metallov i Metalloved., 19: 397 (1965).
785. N. G. Schmahl, Z. Elektrochem., 40: 68 (1934).
786. G. V. Samsonov et al., Electrotechnical Metalloceramic Components, VNIIÉM (1965).
787. O. V. Kamins'ka, Khimichna Promislovist', No. 3, p. 62 (1962).
788. J. S. Fritz et al., Analyt. Chem., 30: 1111 (1958).
789. M. A. Damiens, Bull. Soc. Chim. France, 15: 372 (1914).
790. M. A. Damiens, Ann. Chim., 9: 137 (1918).
791. J. Schmidt, Z. Elektrochem., 40: 170 (1934).
792. M. A. Gurevich et al., Zh. Anal. Khim., 11: 177 (1956).
793. W. Hillebrand and G. E. F. Lundell, Applied Inorganic Analysis, Wiley, New York (1958).
794. A. M. Dymov, Technical Analysis of Ores and Metals, Metallurgizdat (1948).
795. V. I. Smirnov and B. F. Ormont, Zh. Anal. Khim., 9: 359 (1954).
796. B. F. Ormont and B. I. Smirnov, Zh. Fiz. Khim., 31: 281 (1957).
797. N. M. Popova and K. P. Sorokina, Zavod. Lab., 21: 527 (1955).
798. N. M. Popova and K. P. Sorokina, Zavod. Lab., 19: 1033 (1953).
799. O. Meinhardt and O. Krisement, Z. Naturforsch., 15a: 880 (1960).
800. J. A. Leahe, Metallurgia, 45: 989 (1952).
801. J. D. Kennedy, Steel, 131(5): 92 (1952).
802. T. Ya. Kosolapova and G. V. Samsonov, Zh. Fiz. Khim., 35: 363 (1961).

803. J. Hinnüber and O. Rüdiger, Arch. Eisenhuettenwes., 24:267 (1953).
804. A. Westgren et al., J. Iron Steel Inst., 117:386 (1928).
805. T. Ya. Kosolapova and G. V. Samsonov, Proceedings of the Seminar on Heat-Resistant Materials, Izd. Akad. Nauk UkrSSR, Kiev (1961).
806. T. Ya. Kosolapova and G. V. Samsonov, Ukrainsk. Khim. Zh., 28:931 (1962).
807. T. Ya. Kosolapova and S. V. Radzikovskaya, Zavod. Lab., 26:138 (1960).
808. V. Dufek and Z. March, Hutnicke Listy, 14:909 (1959).
809. W. Stolper, Neue Huette, No. 2, p. 110 (1964).
810. I. M. Kolthoff and E. B. Sandell, Textbook of Quantitative Inorganic Analysis, Macmillan, New York (1931).
811. H. Boosz, Z. Anorg. Chem., 292:236 (1957).
812. T. N. Nazarchuk and L. E. Pechentkovskaya, Zavod. Lab., 27:256 (1961).
813. A. Newkirk, J. Am. Chem. Soc., 77:4521 (1955).
814. W. Webb et al., J. Electrochem. Soc., 103:112 (1956).
815. L. A. Mashkovich et al., Zavod. Lab., 30(5):522 (1964).
816. G. Singh and A. R. Verma, Acta Cryst., 17(1):49 (1964).
817. P. Krishna and A. R. Verma, Acta Cryst., 17(2):51 (1964).
818. A. H. Gomesde et al., Acta Cryst., 18:128 (1964).
819. B. Jeantet and A. G. Knapton, Planseeber. Pulvermet., 12:12 (1964).
820. E. Dancy et al., Trans. Met. Soc. AIME, 224:1095 (1962).
821. E. N. Shchetnikov and G. P. Shveikin, Poroshkovaya Met., No. 4, p. 1 (1965).
822. T. Ya. Kosolapova and G. V. Samsonov, Dokl. Akad. Nauk UkrSSR, No. 3, p. 298 (1959).
823. É. G. Airapetyants and E. A. Kachalova, Zavod. Lab., 31:412 (1965).
824. N. A. Suvorovskaya et al., Technological Analysis in Nonferrous Metallurgy, Metallurgizdat (1957).
825. E. B. Sandell, Colorimetric Determination of Traces of Metals (Chemical Analysis, Vol. 3), Interscience, New York (1959).
826. R. Prshibil, Complexones in Chemical Analysis, Goskhimizdat (1960).
827. J. Schmidt, Z. Anorg. Chem., 216:85 (1933).
828. G. V. Kukolev, The Chemistry of Silicon and the Physical Chemistry of Silicates, Gosstroiizdat (1951).
829. A. I. Miklashevskii, Zavod. Lab., 7(2):168 (1938).
830. K. S. Evstrop'ev and N. A. Toropov, The Chemistry of Silicon and the Physical Chemistry of Silicates, Gosstroiizdat (1956).
831. I. S. Kainarskii and É.V. Degtyareva, Carborundum Refractories, Metallurgizdat (1963).
832. A. I. Miklashevskii, Carborundum, Its Chemical Analysis and Properties, GONTI (1938).
833. T. Ya. Kosolapova and T. E. Kotlyar, Zavod. Lab., 24(12):1442 (1958).
834. H. J. Boosz, Techn.-Wiss. Abhandl. Osram Ges., Sonderdruck, 8B:173 (1963).
835. Silicon Carbide, Izd. "Naukova Dumka," Kiev (1966).
836. G. V. Samsonov and V. V. Stasovskaya, Poroshkovaya Met., No. 12, p. 95 (1966).
837. O. Mühlhauser, Z. Anorg. Chem., 5:105 (1894).
838. H. Funk and R. Schauer, Chem. Techn., 6:432 (1954).

839. P. L. Norton, Analyt. Chem., 25:1761 (1953).
840. H. W. Ardelt et al., Z. Chem., 3:70 (1963).
841. A. N. Shteinberg, Tr. Inst. Met. Akad. Nauk SSSR, 11:243 (1962).
842. G. Ehrlich and R. Gerdatsch., Chemia Analiticzna (Polska), 7(2):435 (1962).
843. S. N. L'vov et al., Poroshkovaya Met., No. 4, p. 20 (1962).
844. Ya. I. Gerasimov et al., Chemical Thermodynamics in Nonferrous Metallurgy, Vol. 3, Metallurgizdat (1963).
845. É. G. Airapetyants et al., Zh. Priklad. Khim., 40(2):443 (1967).
846. N. M. Popova and L. V. Zaslavskaya, Zavod. Lab., 21: 1285 (1955).
847. I. G. Shafran and M. V. Pavlova, Zavod. Lab., 7(11):1245 (1938).
848. T. N. Nazarchuk, Zh. Neorgan. Khim., 4(12):2665 (1959).
849. T. N. Nazarchuk, Proceedings of a Seminar on Heat-Resistant Materials, Izd. Akad. Nauk UkrSSR, Kiev (1961).
850. G. V. Samsonov and O. I. Kopnina, Zavod. Lab., 19:269 (1953).
851. A. M. Rekov and G. V. Samsonov, Poroshkovaya Met., No. 5, p. 82 (1961).
852. E. Smith, Analyt. Chim. Acta, 13:15 (1955).
853. T. N. Nazarchuk and L. N. Mekhanoshina, Poroshkovaya Met., No. 2, p. 46 (1964).
854. G. A. Meerson and G. V. Samsonov, Zavod. Lab., 16:1423 (1950).
855. T. Takashi et al., J. Phys. Soc. Japan, 16:2453 (1961).
856. J. P. Bouchaud and R. Fruchart, Bull. Soc. Chim. France, No. 275, p. 1579 (1964).
857. J. P. Bouchaud and R. Fruchart, Compt. Rend., 259:160 (1964).
858. H. Bittner and H. Goretzki, Monatsh. Chem., 91:616 (1960).
859. R. V. Sara, J. Am. Ceram. Soc., 48(6):243 (1965).
860. J. Gurland, J. Metals, 9(4):152 (1957).
861. I. S. Kainarskii et al., Ogneupory, No. 4, p. 175 (1960).
862. A. A. Vertman et al., Dokl. Akad. Nauk SSSR, 159:121 (1964).
863. S. Miczeyewska and T. Niemyski, J. Less-Common Metals, 8:368 (1965).
864. P. Costa and R. R. Conte, Compounds of Interest in Nuclear Reactor Technology (I. Waber and P. Chiotti, eds.), The Metals Society AIME, Michigan (1964), p. 784.
865. O. A. Esin and P. V. Gel'd, Zh. Priklad. Khim., 31:1285 (1958).
866. W. W. Fesenco et al., Colloque Internat. Les Propriétés Méchaniques et Physico-Chimiques des Réfractaires à Haute Témperature, Paris (1965).
867. A. A. Vertman et al., Dokl. Akad. Nauk SSSR, 162(6):1304 (1965).
868. N. J. Grant and O. Preston, J. Metals, 9:349 (1957).
869. D. M. Skorov (ed.), The Materials Science of Reactor Materials, Vol. 3, Atomizdat (1962).
870. E. Dancy et al., Trans. Met. Soc., AIME, 224:1095 (1962).
871. N. D. Morgulis, Usp. Fiz. Nauk, 43:536 (1964).
872. J. R. Tinklepaugh and W. B. Crandall (eds.), Cermets, Reinhold, New York (1960).
873. H. R. Ogden et al., Trans. Met. Soc. AIME, 227:1458 (1963).
874. M. Münster and W. Ruppert, Z. Elektrochem., 57:564 (1953).
875. H. Hamijan and W. Lidman, J. Metals, 5(5):696 (1953).
876. V. V. Pen'kovskii and G. V. Samsonov, Avtomat. Svarka, No. 2, p. 39 (1962).

877. G. V. Samsonov and P. S. Kislyi, High-Temperature Nonmetallic Thermocouples and Terminals, Izd. "Naukova Dumka," Kiev (1965).
878. V. A. Kirillin and A. E. Sheidlin (eds.), High-Temperature Investigations [Russian translation], IL (1962).
879. S. P. Rakitin et al., Teplofiz. Vysokikh Temperatur, 1:145 (1963).
880. N. D. Morgulis, Usp. Fiz. Nauk, 70:679 (1960).
881. L. M. Litz, High-Temperature Investigations [Russian translation], IL (1962).
882. M. Petrdlick and V. Dufex, Neue Huette, 3:1 (1958).
883. S. Peek, Illum. Engr., 52:96 (1957).
884. G. L. Bacchela et al., Bull. Soc. Trans. Miner. Crist., 89:226 (1966).
885. V. V. Grigor'eva et al., Production Technology and Field of Application of Chromium Carbide Alloys, TsITÉIN No. M-60-207/3 (1960).
886. I. Kennedy, Materials and Methods, 38:166 (1952).
887. V. V. Grigor'eva and V. M. Klimenko, Alloys Based on Chromium Carbide, Izv. Akad. Nauk UkrSSR, Kiev (1961).
888. Yu. B. Paderno et al., Élektronika, No. 4:165 (1959).
889. G. A. Gaziev et al., Dokl. Akad. Nauk SSSR, 140 (4): 863 (1961).
890. V. S. Rakovskii et al., The Production of Hard Alloys, Metallurgizdat (1960).
891. F. W. Glaser, Planseeber. Pulvermet., 2:59 (1954).
892. A. I. Avgustinik et al., Izv. Akad. Nauk SSSR, Neorgan. Mat., 1(6):830 (1965).
893. A. Heiss and M. Dodé, Colloque Internat. Les Propriétés Méchaniques et Physico-Chimiques des Réfractaires à Haute Température, Paris (1965).
894. E. E. Kotlyar and T. N. Nazarchuk, Izv. Akad. Nauk SSSR, Neorgan. Mat., 2(10):1778 (1966).
895. A. R. Ubbelohde and F. A. Lewis, Graphite and Its Crystalline Compounds, Oxford University Press, New York (1960).
896. G. B. Bokii, Introduction to Crystallochemistry, Izd. MGU (1954).
897. B. Aronson, Acta Chem. Scand., 13:100 (1959).
898. G. S. Zhdanov and N. G. Sevast'yanov, Zh. Fiz. Khim., 17:326 (1943).
899. A. L. Burykina and L. V. Strashinskaya, in: High-Temperature Inorganic Compounds, Izd. "Naukova Dumka," Kiev (1965).
900. M. Kh. Karapetyants and M. L. Karapetyants, Tables of Some Thermodynamic Properties of Various Substances, Izd. MKhTI im. Mendeleeva (1961).
901. B. W. Roberts, Superconductive Materials and Some of Their Properties, New York (1963).
902. S. N. L'vov and V. F. Nemchenko, in: High-Temperature Inorganic Compounds, Izd. "Naukova Dumka," Kiev (1965).
903. G. V. Samsonov and V. S. Sinel'nikova, Poroshkovaya Met., No. 4, p. 59 (1962).
904. R. C. Vickery, J. Chem. Soc. (1959), p. 503.
905. G. V. Samsonov, in: Methods and Apparatus for Microhardness Testing, Izd. "Nauka" (1965).
906. G. V. Samsonov and V. S. Neshpor, in: Problems in Powder Metallurgy and Strength of Materials, No. 5, Izd. Akad. Nauk UkrSSR (1958).
907. B. R. Lazarenko, Vestn. Akad. Nauk SSSR, 4:83 (1961).
908. V. K. Kharchenko and L. I. Struk, Poroshkovaya Met., No. 2, p. 87 (1962).
909. G. V. Samsonov and V. S. Sinel'nikova, Ukrainsk. Fiz. Zh., 6(5):687 (1961).

910. G. V. Samsonov et al., Ukrainsk. Fiz. Zh., 10:622 (1965).
911. J. Besson et al., Compt. Rend., 261:1859 (1965).
912. H. Schenk et al., Forschungsber. Landes Nordrein-Westfalen, No. 1589, p. 81 (1966).
913. V. V. Pen'kovskii, Poroshkovaya Met., No. 5, p. 74 (1961).
914. E. E. Kotlyar and T. N. Nazarchuk, Zh. Analit. Khim., 15(2):207 (1960).
915. J. J. Harwood, Metallurgiya, 47:43 (1953).
916. F. P. Knudsen et al., J. Am. Ceram. Soc., 38:312 (1955).
917. L. Yakeshova, New Nuclear Materials Including Nonmetallic Fuels, Vol. 1, International Atomic Energy Agency, Vienna (1969), p. 3771.
918. T. Ya. Kosolapova et al., Izv. Akad. Nauk SSSR, Neorgan. Mat., 2(8):1516 (1966).
919. L. I. Struk, Izv. Akad. Nauk SSSR, Neorgan. Mat., 3(4):624 (1967).
920. N. H. Krikorian et al., J. Nucl. Mater., 21(2):236 (1967).
921. R. Juza et al., Z. Anorg. Chem., 352:89 (1967).
922. S. S. Ordan'yan and A. I. Avgustinik, Izv. Akad. Nauk SSSR, Neorgan. Mat., 4:1898 (1968).
923. S. S. Ordan'yan et al., Izv. Akad. Nauk SSSR, Neorgan. Mat., 4:1086 (1968).
924. Yu. B. Paderno et al., Izv. Akad. Nauk SSSR, Neorgan. Mat., 5:889 (1969).
925. S. Saburo and S. Yoshinobu, J. Japan. Inst. Metals, 21:271 (1957).
926. R. Steinitz and R. Resnick, J. Appl. Phys., 37:3463 (1966).
927. J. P. Bouchand, Ann. Chim., 2:353 (1967).
928. H. Nickel et al., Z. Metallkunde, 59:935 (1968).
929. N. I. Krasotkina et al., Ogneupory, No. 11, p. 49 (1968).
930. R. Lallement et al., Compt. Rend., 259:2442 (1964).
931. F. A. Josix and R. Renaud, Compt. Rend., 260:2239 (1965).
932. Babod-Zakhryapin et al., Izv. Akad. Nauk SSSR, Neorgan. Mat., 5:1133 (1969).
933. R. H. Flowers and E. G. Rauh, J. Inorg. Nucl. Chem., 28:1355 (1966).
934. M. J. Philippot, Compt. Rend., 264(Sec. C):1877 (1967).
935. J. P. Hammond and G. M. Adamson, Jr., Modern Developments in Powder Metallurgy (H. Hausner, ed.), Vol. 3, Plenum Press, New York (1966).
936. Zorenzelli, Compt. Rend., Vol. 266 (1968).
937. G. V. Samsonov et al., Poroshkovaya Met., No. 11, p. 42 (1968).
938. G. V. Samsonov et al., Poroshkovaya Met., No. 8, p. 107 (1966).
939. G. V. Samsonov et al., Poroshkovaya Met., No. 4, p. 22 (1928).
940. A. S. Volgar et al., Poroshkovaya Met., No. 4, p. 60 (1968).
941. G. G. Gnesin and Yu. V. Naidich, Poroshkovaya Met., No. 4, p. 57 (1969).
942. V. K. Vitryanyuk and V. V. Ordenko, Poroshkovaya Met., No. 4, p. 34 (1967).
943. C. E. Ryan et al., Anisotropy in Single-Crystal Refractory Compounds, Vol. 1, (1968), p. 177.
944. R. Kieffer et al., Ber. Deut. Keram. Ges., 43:621 (1966).
945. J. C. Vignie and J. Spitz, Rapp. CEA-R. 3155 (1967).
946. M. M. R. Fruchart et al., Compt. Rend., 266(Sec. C):1961 (1968).
947. J. P. Senateur, Ann. Chim., 2:103 (1967).
948. G. V. Samsonov and G. Sh. Upadkhaya, Poroshkovaya Met., No. 5, p. 69 (1969).
949. A. T. Turchanin and V. V. Fesenko, Poroshkovaya Met., No. 6, p. 48 (1969).

950. V. S. Telegus et al., Kristallografiya, 12: 936 (1967).
951. J. M. Leintaker and T. G. Godfrey, J. Nucl. Mat., 21: 175 (1967).
952. T. Satow, J. Nucl. Mat., 21: 249, 255 (1967).
953. P. Chiotti et al., J. Nucl. Mater., 23: 55 (1967).
954. R. Kieffer and H. Rassaerts, Internat. J. Powder Met., 2(2): 15 (1966).

Index*

*T or F listed after a page reference indicates that the subject is to be found in a table or figure, respectively.

MIX
Papier aus verantwortungsvollen Quellen
Paper from responsible sources
FSC® C105338

If you have any concerns about our products,
you can contact us on
ProductSafety@springernature.com

In case Publisher is established outside the EU,
the EU authorized representative is:
Springer Nature Customer Service Center GmbH
Europaplatz 3, 69115 Heidelberg, Germany

Printed by Libri Plureos GmbH
in Hamburg, Germany